WHELDON'S
BUSINESS STATISTICS
AND STATISTICAL METHOD

WHELDON'S
BUSINESS STATISTICS
AND STATISTICAL METHOD

G.L. THIRKETTLE,
B. Com.(Lond.), F.I.S.

Formerly Senior Lecturer in Statistics
The Polytechnic of North London

NINTH EDITION

MACDONALD AND EVANS

MACDONALD & EVANS LTD.
Estover, Plymouth PL6 7PZ

First published 1936
Second edition 1940
Third edition 1945
Reprinted 1945
Reprinted 1946
Reprinted 1947
Reprinted 1948
Reprinted 1951
Reprinted 1952
Reprinted 1953
Reprinted 1955
Fourth edition 1957
Reprinted 1958
Reprinted 1960
Reprinted 1961
Reprinted 1962
Fifth edition 1962
Reprinted 1963
Reprinted 1964
Reprinted (with amendments) 1965
Reprinted 1966
Reprinted 1967
Sixth edition 1968
Reprinted 1969
Seventh edition 1972
Eighth edition 1976
Reprinted 1978
Ninth edition 1981
Reprinted 1985

© Macdonald & Evans Ltd, 1981

0 7121 2324 5

Printed in Great Britain by
The Pitman Press, Bath

Preface to the Ninth Edition

Competent administration, whether in commerce, industry or government, requires a knowledge of statistical method. This fact has become more and more recognised, as is evidenced by the large number of bodies that now include the subject of Statistics in their examinations.

There is also an ever-growing number of people in business who, whilst having no examination in view, are finding it more than useful to have a knowledge of statistics. Since this book avoids a purely theoretical approach and places emphasis on applications to business problems, it should be of great help to them.

So that this book should continue to serve students in the future as it has done so well in the past, this new up-dated edition has been extensively revised and a number of chapters entirely rewritten.

The main changes are :-

(a) much new material;

(b) a series of programmes on statistical procedures — many students find it useful to have a list of things to do when solving a statistical problem;

(c) the addition of many warning *Notes* where experience has shown students are prone to error;

(d) many new diagrams specially designed to clarify the accompanying text.

I thank all those bodies (a list of which appears at the beginning of *Appendix I*) who have kindly allowed me to reproduce questions from their examinations. Any solutions I have given are solely my responsibility.

Acknowledgments are also due to the Controller of HMSO for permission to reproduce copyright material from a number of government publications.

1980   G.L.T.

Contents

two sets of control limits. Fraction defective charts. Cusum charts.

Appendixes

CHAPTER 1

Introduction

THE MEANINGS OF "STATISTICS"

This word is used with two quite distinct meanings. It can refer to facts which can be put into a numerical form, as in the phrase "unemployment statistics". This is the meaning the man in the street gives to the word. It can also refer to statistical methods, which is the subject-matter of this book. Statistical methods are, to use Bowley's description, devices for abbreviating and classifying numerical statements of facts in any department of inquiry and making clear the relations existing between them.

When used with the first meaning, "statistics" is a plural noun, and the statistician usually uses the word data instead. When "statistics" refers to the science it is a singular noun, and this is the meaning usually given to it by the statistician.

THE NATURE OF STATISTICAL DATA

Statistics can only deal with numerical data. Often, however, data which are of a qualitative nature can be put into a quantitative form. Health can be measured by the number of days' illness; intelligence, as is well known, by specially designed tests. There is a considerable amount of arbitrariness about this, but it does allow statistical methods to be used, and thus provides additional evidence not otherwise available in given inquiries.

Sometimes the material of an inquiry can be divided up into "cases" possessing a certain attribute and those not possessing it. It is often possible to put qualities in order of rank. Thus, in the case of colours, the lightest could be given rank 1, the next lightest rank 2 and so on. In business, however, most data can be measured or

1

counted directly. Examples would be the number of absentees, sales, quantity produced, wages, etc.

THE IMPORTANCE OF DEFINITION

Before any data are collected it is absolutely necessary to define clearly and unambiguously all the terms used and every piece of information required. Failure to do so will mean that the conclusions drawn from the inquiry will not be valid, and that comparisons over time and between areas will not be true comparisons. The conclusions drawn will be inaccurate because answers to questions will not be answers to the same questions, but only apparently so, the questions having many interpretations unless all terms are clearly defined.

STATISTICAL INQUIRIES

The business man, and indeed anyone who has to administer any organisation, is concerned with inquiries of many kinds. Some of these are capable of being treated statistically, and statistical evidence can be provided in respect of the information wanted.

The steps in a statistical inquiry are as follows.

(a) *The problem must be clearly stated.* Suppose the problem concerns wages in a factory. Is it about wages earned or wage rates? Must the statistics concern all employees, or separate grades, both men and women? Should lost time, overtime, piece-work and bonus payments be included or allowed for? Should receipts in kind be included? The purpose of the investigation will provide guidance as to the exact information to be obtained.

(b) *Selection of the sample.* If complete coverage of the information available is not made, then the size of the sample and method of sampling will have to be determined. This will depend on the kind of information wanted, the cost and the degree of accuracy required. The best example of a sample inquiry in business is market research.

(c) *Drafting the questionnaire.* This is quite a difficult job if the answers obtained are to be of value. Usually a number of questions have to be redrafted to get the exact information wanted. A pilot survey is useful to enable a satisfactory questionnaire to be obtained.

A great deal of information in business, however, is already available in the form of accounting records, costings, administrative information about personnel and so on; questionnaires apart from market research are therefore used only for special inquiries.

(d) *Collection of data.* Where not available as administrative records or published, the most satisfactory way is by means of enumerators. Enumerators ask the questions and fill in the questionnaires.

(e) Editing the schedules. Questionnaires require to be checked, sometimes coded, and calculations made before tabulation can be done.

(f) Organisation of data. The items require to be counted or the values summed either in total or in various categories before they can be tabulated.

(g) Analysis and interpretation. Before the information acquired can be used it is analysed and then interpreted. This requires a sound knowledge of statistical methods and also, and this is often lost sight of, a sound knowledge of the subject for which statistical evidence has been obtained.

(h) Presentation. This may take the form of tables, charts and graphs.

(i) The writing of the report. This will give the results of the investigation and, where required, will make recommendations. Tables and charts usually play an important part in business reports.

STATISTICS AS A TOOL OF MANAGEMENT

In order that the student at the beginning of his studies shall have some idea of the value of statistics to the business man, some problems, which are the concern of management and for which statistical methods are appropriate, are given below.

(a) Stock control. Carrying too large a stock means idle capital and unnecessary costs of storage. Too small a stock means that materials are not available when required, resulting perhaps in lost sales. The "right" amount is a matter, therefore, of considerable importance.

(b) Market research. The business man, in order to be successful, requires to sell what his customers want in the right types and qualities. He wants to know what to sell, in what quantities, when and where.

(c) Quality control. If he produces a mass-produced article, the business man can usually check the quality of his output more efficiently by means of statistical methods.

(d) Sales trends. This is vital information for future planning. Statistical methods are preferable to optimistic guesses, usually called judgment.

(e) The relationship between costs and methods of production. Unless such information is forthcoming, the most efficient method of manufacture will probably not be used.

This matter of statistics and business management is returned to in Chapter 24. Before that, statistical method is dealt with in some detail. The student must master the methods before he can apply them.

QUESTIONS

1. "Statistics is the science of counting." Criticise this definition. If you are not satisfied with it, state your own.

2. Why is definition important in any statistical investigation?

3. *(i)* "The more facts one has, the better the judgment one can make."

   (ii) "The more facts one has, the easier it is to put them together wrong."

Comment on these two views in relation to the principle that statistics are an aid to, not a substitute for, business judgment.

Chartered Institute of Transport.

4. Enumerate the main steps in undertaking a statistical inquiry.

5. In what practical ways can statistics be of service to a business man?

CHAPTER 2

Collection of Data

PRIMARY AND SECONDARY DATA

Data may be expressly collected for a specific purpose. Such data are known as primary data. The collection of facts and figures relating to the population in the census provides primary data. The great advantage of such data is that the exact information wanted is obtained. Terms are carefully defined so that, as far as humanly possible, misunderstanding is avoided.

Often, however, data collected for some other purpose, frequently for administrative reasons, may be used. Such data are known as secondary data. Details of imports and exports are compiled by the Statistical Office and Customs and Excise from declarations made by importers and exporters to the local Customs and Excise Officers. This material is obtained for administrative reasons. It is used, however, for compiling quite a large number of statistics relating to overseas trade, including, for example, import and export price indices. The Retail Price Index, on the other hand, uses primary data, the officials of the Department of Employment collecting retail prices expressly for the Retail Price Index each month.

Secondary data must be used with great care. Such data may not give the exact kind of information wanted, and the data may not be in the most suitable form. Great attention must be paid to the precise coverage of all information in the form of secondary data.

Much of the business data used in compiling business statistics is also secondary data, the source often being the accounting, costing, sales and other records. Such data will often require to be adjusted. In finding a sales trend, for example, it may be necessary to adjust the sales figures as given by the accountant for the varying days in the month. Other secondary data used by the business man are

5

published statistics. Before any such material can be used with safety it will be necessary to know the source of the figures, how they were obtained, exact definitions and methods of compilation.

METHODS OF COLLECTING DATA

Business data are often collected in the normal course of administration, and not specifically for statistical purposes. However, there is no reason why records should not serve the two purposes, and in such cases care should be taken to ensure that the record is adequate statistically as well as administratively.

The following list covers the most important methods of collecting data.

(a) Postal questionnaire. This takes the form of a list of questions sent by post. Unless, however, the respondent (the person who is required to answer the questions) has an interest in answering it or is under legal compulsion to do so, the postal questionnaire is generally unsatisfactory, producing few replies, and those of a biased nature.

The postal questionnaire is satisfactory when the law compels the respondent to reply. The Statistics of Trade Act 1947 makes it compulsory for firms to answer the questionnaire sent out in respect of the census of production.

The postal questionnaire is also satisfactory when sent by trade associations to their members, since the members have an interest in answering it.

Some firms have tried to get answers by offering small gifts. This is not a very good idea, since it will produce biased answers, because the respondent tries to please the donor.

(b) Questionnaires to be filled in by enumerators. This is the most satisfactory method. The enumerators or field-workers can be briefed so that they understand exactly what the questions mean; they can get the "right" answers; and they fill in the questionnaires more accurately than would the respondents themselves.

In the case of the Census of Population of the UK, the enumerators do not fill in the schedules. This, however, is quite exceptional.

(c) Telephone. Asking questions by telephone is not usually a very good method, because people who possess telephones form a biased sample. Telephone interviews are, however, useful for certain kinds of radio research.

(d) Observation. This method entails sending observers to record what actually happens while it is happening. An example where this method is suitable is in the case of traffic censuses.

Actual measurements or counting also come under the heading of observations. Examples occur in Statistical Quality Control.

(e) Reports. These may be based on observations or informal conversations. They are usually incomplete and biased, but in certain cases may be useful.

(f) Results of experiments. This method is of greater interest to the production engineer, the agronomist and other applied scientists than to the business man.

STATISTICAL UNITS

Since the compilation of statistics necessarily entails counting or measurement, it is very important to define the statistical unit which is to be counted or measured.

The definition of the unit is not always as simple as would at first appear. A little reflection on the following words will reveal the need for precise specification in each case: prices may be wholesale, retail, delivered or ex-works; accident may refer to a slight or serious injury, one officially reported or one resulting in a compensation claim; wages may refer to earnings or wage rates.

REQUIREMENTS FOR A STATISTICAL UNIT

The requirements for a statistical unit are as follows.

(a) It must be clearly defined and unambiguous.

(b) It must be homogeneous. This uniformity is essential; the unit must not imply different characteristics at different times and places. If the selected unit is not applicable to all cases coming under review, it is often possible to overcome the difficulty *(i)* by subdividing the data into groups or classes until sufficient uniformity has been secured, *(ii)* by expressing dissimilar units in terms of equivalents of the selected unit. For example, if the output of bakeries for a period were being compared, some producing 1-, 2- and 4-kg loaves, all sizes could be reduced to the equivalent of, say, 2-kg loaves.

(c) It should be stable. If it is desired to use a fluctuating unit, such as a calendar month, then adjustments will require to be made before comparisons will be valid.

(d) It must be appropriate to the inquiry and capable of correct

ascertainment. When compiling labour statistics it is necessary to select the unit appropriate to the information required, e.g. workers engaged directly on production, those employed in indirect factory services, those in administrative offices or those in the sales department.

TYPES OF STATISTICAL UNIT

A simple statistical unit—e.g. metres, litres, £ sterling—are not difficult to define, but care must be used in some cases. For example, a ream may be 480 sheets, but often consists of 500 or 516 sheets. A ton may be 2,240 lb or it may be a short ton of 2,000 lb, or again it may be a metric ton (tonne).

A composite unit may have to be used in some cases. Thus, electric power is measured in units of kilowatt-hours. Education authorities measure the amount of instruction in student-hours.

QUESTIONNAIRES

THE FORM OF QUESTIONNAIRE

The questionnaire is often in two parts. The first part is a classification section. This requires such details of the respondent as sex, age, marital status, occupation. The second part has the questions relating to the subject-matter of the inquiry. The answers given in the second part can be analysed according to the information in the first part. There will also be questions of a purely administrative nature, such as the date of the interview, the name of the interviewer, etc. Market research and social survey questionnaires are usually of this form.

Concerning the layout of questionnaires, there is a tendency, where it is possible, to provide all the answers that could be given to a question on the form, and the appropriate answer is circled or ticked. This treatment can only, of course, apply to certain questions.

THE CHARACTERISTICS OF A GOOD QUESTIONNAIRE

The list of desirable qualities that a questionnaire should possess given below would seem to be a matter of common sense. Nevertheless, the drafting of questionnaires is one of the most difficult tasks of an inquiry. A pilot survey—that is, a trial survey, carried out prior to the actual survey—invariably leads to alterations and improvements in the questionnaires.

(a) *Questions should not be ambiguous.* This means that the questions must be capable of only one interpretation.

(b) *Questions must be easily understood.* Technical terms should be avoided, except where the questionnaire is addressed to specialists.

(c) *Questions should be capable of having a precise answer.* The answer should take the form of "yes" or "no", a number, a measurement, a quantity, a date, a place; facts are required, not opinions (except where opinions are wanted, as in opinion polls).

(d) *Questions must not contain words of vague meaning.* To ask if something is large or if a man is unskilled are examples of such questions. When does something become large? What jobs are unskilled?

(e) *Questions should not require calculations to be made.* Such questions give rise to unnecessary sources of error. If for the purpose of the inquiry it is necessary to know annual earnings, but the respondent is paid weekly, the weekly earnings are asked for. The calculations necessary are done by the statistician's staff.

(f) *Questions should not require the respondent to decide upon classification.*

(g) *Questions must not be in such a form that the answers will be biased.* The questions will not therefore contain emotionally coloured words, they will not be leading questions—that is, they will not put answers into the respondents' mouths—and, of course, they must not give offence.

(h) *The questionnaire should not be too long.* If a questionnaire is too long, the respondent will not be co-operative, and this may mean inaccurate answers; often a way out is for some of the respondents to answer part of the questions and other respondents to answer other questions.

(i) *The questionnaire should cover the exact object of the inquiry.* However, provided the questionnaire is not made too long, advantage can be taken of the arrangements made to obtain information relating to a different inquiry. In one population census of the UK, questions were asked about sanitary arrangements.

When drafting questionnaires, it is a good plan to check the questions against the above list.

INTERVIEWING

The enumerator does not usually deliver the questionnaires, but, instead, asks the respondent the questions listed. He, or she (and the majority of enumerators are women), also fills in the answers. Upon the enumerator, then, falls the duty of obtaining accurate information. The interviewer must not express his own opinions, nor must he allow them to influence the answers. Complete objectivity must be observed. The interview must merely record facts accurately—that is, the answers as given.

The enumerator should first make sure he is interviewing the right person. He should then explain the purpose of the interview as clearly and as briefly as possible, soliciting the help of the respondent. If help is refused, further details about the inquiry may succeed in obtaining the required co-operation.

If the respondent does not understand the question, or has not given a clear answer, resort is made to "probing". This consists of re-wording or explaining the question, or sometimes asking other questions to help a person to remember. It also consists of asking a respondent to explain his answers more fully. Great care must be taken not to put answers into the respondent's mouth. Where the question asks for an opinion, all the probing allowed to the enumerator is to ask the respondent to explain an unclear answer more fully. He must not explain or re-word questions.

NON-RESPONSE

Unless every person to be interviewed is interviewed the results will not be valid. Non-response must therefore be kept to negligible proportions. If a respondent is not interviewed when first called upon, it will be necessary to follow up with a second call and not to substitute someone else.

If there is a good response to the follow up, the initial non-respondents who subsequently respond can be treated as a sample of all the initial non-respondents and weighted accordingly. Consider a case where 2,000 people are to be interviewed to find out the proportion of people taking holidays abroad. Of these people, 600 were out when the interviewer called, but of the other 1,400, 420 took their holidays abroad, that is 30 per cent. Of the 600 non-respondents 200 responded to a follow up and of these 80 had holidays abroad, that is 40 per cent. The 200 subsequent respondents are considered to be typical of all 600 initial non-respondents; the whole of the 600 are therefore considered to have 40 per cent of their number having holidays abroad. The combined result is:

$$\frac{30\% \text{ of } 1{,}400 + 40\% \text{ of } 600}{2{,}000} = 33\%.$$

QUESTIONS

1. Give some account of the various methods which can be used in the collection of statistical data.

2. "The collection of accurate statistical data upon a large scale is only possible with exact definition of terms, e.g. units, classes, etc." Comment with illustrations from the transport industry.

Chartered Institute of Transport.

3. Design a questionnaire to investigate the amount spent on clothing and footwear throughout the country. State the precautions you would take to avoid any bias in your results.

London Chamber of Commerce and Industry.

4. By drawing on your own or related experience in the transport industry, illustrate the need for, and importance of, exact definition for statistical quantities.

Chartered Institute of Transport.

5. List the main points to be observed when drafting a questionnaire.

6. Name four ways in which statistics can be of service to a manufacturer.

7. Distinguish between primary and secondary data, giving examples of each.

8. What care must be taken when using secondary data?

9. What is meant by probing? When is it permissible?

CHAPTER 3

Sampling

THE SAMPLING PROCESS

Instead of obtaining data from the whole of the material being investigated, sampling methods are often used in which only a sample selected from the whole is dealt with, and from this sample conclusions are drawn relating to the whole. If the conclusions are to be valid, the sample must be representative of the whole. The selection of this sample must therefore be made with great care.

THE ADVANTAGES OF USING SAMPLING METHODS

There are a number of advantages to be gained by using sampling methods. Among these may be listed the following.

(a) Quick results. The results can be obtained more quickly, for two reasons: the data are obtained more rapidly and can be analysed more quickly. Speed in analysing the data is also possible by analysing a sample of the data before analysing the whole. The analysis of a 1 per cent sample of the 1951 population census was such an example.

(b) Higher quality interviewers. When the respondents constitute only a small proportion of the population they are more willing to give detailed information, and because fewer interviewers are required, a much higher quality of interviewer can be employed.

(c) More skilled analysis. Higher-grade labour can be employed on the computing, tabulating and analysis of the data.

(d) Lower costs.

(e) Following up non-response is much easier.

(f) The error can be assessed.

(g) Guarding against incomplete and inaccurate returns is easier. A sample is often used as a check on the accuracy of complete censuses.

(h) Practical method. When the investigation entails the destruction of the material—e.g. to discover the life of a lamp—sampling methods are the only practical methods to use.

THE "POPULATION" OR "UNIVERSE"

These terms refer to the whole of the material from which the sample is taken. The frame will consist of all the items in the population, or some means of identifying any particular item in the population. This frame is necessary so that any item in the population can be part of the sample. The frame must be complete—that is, no item of the population should be left out—and it should not be defective because it is out of date or contains inaccurate or duplicate items, or inadequate because it does not cover all the categories required to be included in the investigation.

Examples of frames suitable for certain inquiries are the electoral lists. For the sampling of dwellings in built-up areas, town maps provide a useful frame.

Sometmes, no suitable frame is available, the population required may be part of a larger population, and cannot be identified within it. One method that can then be used is to make the first question determine if the respondent is within the special population required. This method entails a good deal of work to get the necessary sample.

THE POSTCODE SYSTEM AS A SAMPLING FRAME

For market research and many economic and sociological surveys where sampling is on an area basis, postcodes provide an extremely useful frame.

One and a half million postcodes cover twenty-two million addresses in the UK. Some addresses (those having large quantities of mail) have their own postcodes (large-user postcodes). The other addresses are covered by small-user postcodes (an average of 15 addresses to each).

The postcode consists of up to seven letters and numbers which denote four levels of area. The largest area is a postcode area denoted by the first one or two letters. Each postcode area is divided into postcode districts denoted by the figures in the first part of the code. The districts are divided into postcode sectors denoted by the figure that occurs in the second part of the postcode. The last two letters pinpoint one street or part of a street.

There are maps available which show postcode area boundaries and postcode district boundaries.

THE BASIC STATISTICAL LAWS

The possibility of reaching valid conclusions concerning a population from a sample is based on two general laws, namely, the law of statstatistical regularity and the law of the inertia of large numbers.

(a) The law of statistical regularity. The law of statistical regularity states that a reasonably large number of items selected at random from a large group of items will, on the average, be representative of the characteristics of the large group (or "population"). The important features are as follows.

(i) The selection of samples must be made at random, as for instance by some lottery method, so that every item in the population has a chance of being in the sample.

(ii) The number of items in the sample must be large enough to avoid the undue influence on the average of abnormal items. The larger the number of items selected, the more reliable is the information afforded.

(b) The law of inertia of large numbers. The law of inertia of large numbers is a corollary of the law just described. It states that large groups or aggregates of data show a higher degree of stability than small ones.

The movements of all the separate components of the aggregate reveal a tendency to compensate one another, some probably moving higher, others lower; but, taking a large number of data, it is unlikely they will all move in the same direction. The greater the number composing the aggregate, the easier will be the compensation, or tendency of movements to neutralise one another, and consequently the more stable will be the aggregate.

SAMPLING ERRORS

Even when a sample is chosen in a correct manner, it cannot be exactly representative of the population from which it is chosen, for not all samples will be alike. This gives rise to sampling errors.

The amount of the random sampling error will depend on the size of the sample—the larger the sample, the smaller the error—the sampling procedure involved, and the extent to which the material varies.

In the case of sampling methods which are based on random sampling, it is possible to measure the probability of errors of any given size. By error is meant of course, the difference between the estimate of a value as obtained from the sample and the actual value.

ERRORS DUE TO BIAS

In addition to random sampling errors, errors due to bias may arise. Unlike sampling errors, errors due to bias increase with the size of the sample. Furthermore, it is not possible to measure the amount. Hence, every care must be taken to avoid these kinds of errors. They may arise in a number of ways.

(a) *Deliberate selection.* Whenever the personal element enters into the selection, bias is inevitably introduced.

(b) *Substitution.* Substituting another item for one already chosen in the sample will also cause bias. This may happen if, for example, a person chosen for interviewing is out, and someone else is then interviewed. People who are at home all day have not the same characteristics as those who spend most of their time away from home.

(c) *Failure to cover the whole of the sample.* Even if no attempt is made to substitute and the whole sample is not covered, there will be bias. This is a frequent fault with postal questionnaires, when not all the respondents reply.

(d) *Haphazard selection.*

TYPES OF SAMPLING

RANDOM SAMPLING

The word random does not mean haphazard. It refers to a definite method of selection. In a *simple random sample* each unit of the population has exactly the same chance as any other unit of being included in the sample. One method of obtaining a random sample, not a very practical one, would be to number or name every unit of the population, and to place slips, each bearing a number or a name, one slip for each unit of the whole of the population, in a drum, as in a lottery, and, after thoroughly mixing them, draw from the drum the number of slips required to obtain the required size of sample. In practice, a table of random numbers is used. A small portion of such a table is shown in Fig. 1. These numbers can be read in any methodical way: vertically, horizontally, diagonally, forwards or backwards, etc. Any number chosen must be capable of being assigned unambiguously to a particular unit of the population.

Suppose the population to consist of 999,999 items, the sample required 1,000, then the first four items to be chosen in the sample might be, using the table shown, numbered 162,277; 844,217; 630,163; 332,112. The figures have been read across for each number, but downwards to get the next number, starting with the second group of numbers in the left-hand column. Countless alternatives could have been chosen.

03	47	43	73·	86		36	96	47	36	61		46	98	63	71	62
97	74	24	67	62		42	81	14	57	20		42	53	32	37	32
16	76	62	27	66		56	50	26	71	07		32	90	79	78	53
12	56	85	99	26		96	96	68	27	31		05	03	72	93	15
55	59	56	35	64		38	54	82	46	22		31	62	43	09	90
16	22	77	94	39		49	54	43	54	82		17	37	93	23	78
84	42	17	53	31		57	24	55	06	88		77	04	74	47	67
63	01	63	78	59		16	95	55	67	19		98	10	50	71	75
33	21	12	34	29		78	64	56	07	82		52	42	07	44	38
57	60	86	32	44		09	47	27	96	54		49	17	46	09	62
18	18	07	92	46		44	17	16	58	09		79	83	86	19	62
26	62	38	97	75		84	16	07	44	99		83	11	46	32	24
23	42	40	64	74		82	97	77	77	81		07	45	32	14	08
52	36	28	19	95		50	92	26	11	97		00	56	76	31	38
37	85	94	35	12		83	39	50	08	30		42	34	07	96	88
70	29	17	12	13		40	33	20	38	26		13	89	51	03	74
56	62	18	37	35		96	83	50	87	75		97	12	25	93	47
99	49	57	22	77		88	42	95	45	72		16	64	36	16	00
16	08	15	04	72		33	27	14	34	09		45	59	34	68	49
31	16	93	32	43		50	27	89	97	19		20	15	37	00	49

Fig. 1. Small portion of a table of random numbers.

STRATIFIED SAMPLE

Where the population is heterogeneous, a stratified sample is required. This will increase the accuracy of the results, provided the strata relevant to the investigation are chosen. The population is divided into strata, blocks of units, in such a way that each block is as homogeneous as possible. Each block or stratum is then sampled at random. If the same proportion of each stratum is taken, each stratum will be represented in the correct proportion in the sample. This eliminates differences between strata from the sampling error.

Sometimes instead of a *uniform sampling fraction, variable sampling fractions* are employed. The more variable strata in the population is more intensively sampled. This will increase the precision.

SYSTEMATIC SAMPLING

When a list of all the units of a population is available a sample of every nth item would be known as a systematic sample. The first entry would be obtained by random selection. However, unless the list is arranged at "random" (in the technical sense), such systematic sampling is not random sampling.

Provided there are no characteristics of the items in the list which occur periodically with the same interval as the sampling interval, the sample will not be biased. It is also possible to obtain an estimate of the sampling error good enough for most practical purposes.

MULTI-STAGE SAMPLING

This is best explained by example. Instead of obtaining a sample from, say, all the households in the country, a random sample is obtained of all the rating authorities. A random sample would then be chosen of the households in each of the areas of the rating authorities chosen at the first stage of sampling.

QUOTA SAMPLING

The sample is divided up into quotas, the quotas indicating the number of people to be interviewed, but leaving the choice of the actual respondents to the interviewers. This, of course, introduces bias. The quotas are chosen so that the sample is representative of the population in a number of respects, according to the controls chosen. It may be completely unrepresentative in other respects. Suppose a sample of 400 is required, that the controls chosen are sex, and whether householder or not. Let the proportion of men to women in the population from which the sample is chosen be 9 to 11, and the proportion of householders to non-householders be 3 to 2. The quotas would be as follows: 132 householders who are women, 88 women non-householders, 108 householders who are men and 72 men non-householders. The sample of people interviewed might be completely unrepresentative as regards age. But this might not be relevant for the inquiry. It is not possible to measure the sampling error.

QUESTIONS

1. What is a statistical sample? When and why are samples used? Discuss the essentials of the various methods of sampling.

2. Describe a method of drawing a random sample for a postal survey of annual purchases of toothpaste by households in the

Southern Counties. Design a short questionnaire, including any necessary instructions for completion.

Institute of Statisticians.

3. Explain the following:

Law of statistical regularity.

Law of inertia of large numbers.

What section of statistical method depends on these two laws?

4. Explain the use of sampling in statistical investigations. What considerations should principally be borne in mind when making a social inquiry by sample?

5. What is the object of sampling? Illustrate it by reference to actual cases of which you are aware, referring to some different kinds of sampling and the dangers to be avoided.

Institute of Cost and Management Accountants.

6. Describe the sample method of investigation. Illustrate your answer by reference to any official or non-official statistical investigation with which you are familiar.

7. Describe three of the more important methods of sampling used in conjunction with statistical surveys.

Suggest methods which may be employed in statistical surveys to validate the sample information obtained.

Institute of Chartered Secretaries and Administrators.

8. What is a statistical sample? Discuss the essentials of the various methods of sampling and state and explain the two laws upon which the sampling technique depends.

9. Explain carefully the meaning of the following terms:

(a) population;
(b) frame;
(c) random sample;
(d) quota sampling;
(e) sampling error.

10. In what ways can a faulty selection of the sample give rise to bias?

CHAPTER 4

Surveys

A survey consists in finding facts in particular fields of inquiry. In this chapter, a short description is given of three important types of surveys in which the data collected are of a statistical nature.

SOCIAL SURVEYS

Although many bodies conduct social surveys—as, for example, the universities—the most important are the social surveys of the Government.

The social survey came into being to provide information for other Government departments so that they could carry out their duties more efficiently. Government departments often have to deal with special social problems. The surveys help the Government committees to arrive at a policy for solving the particular problems. A large part of their work consists in providing continuous statistics, such as the Consumer Expenditure inquiries which are published in the *National Income and Expenditure Blue Books.*

The survey is a sample survey, the size of the sample depending on the permissible sampling error and the amount of analysis required (it does *not* depend on the size of the population). The sample may therefore vary from 400 to 4,000, or even more. The sampling is always some form of "random" sampling, the word "random" being used in its technical sense (*see* Chapter 3).

The actual method used is usually some form of multi-stage sampling—e.g. a sample of the 1,470 administrative districts of England and Wales and the fifty-seven administrative districts of Scotland is first taken, and from each of the districts chosen a sample of people will be taken from the card-indexes kept in each of the administrative districts. In the case where a sample of house-holders is required, the rating records kept by local authorities are used. Quota sampling is used only on some pilot surveys when a fully representative sample is not so important, the aim of the pilot survey being to establish the most satisfactory form of the questions to be asked; random sampling is used for the actual survey.

MARKET RESEARCH

Since the aim of the producer is to sell his goods, it is necessary for him to know the "right" kind of commodity to produce, and, in order to prevent losses, to know in what quantities and at what price. This entails the producer knowing the consumers' needs and desires, and if he attempts to persuade them of the superiority of his particular products, knowing the results of his publicity and his competitors' successes. This part of market research lends itself admirably to statistical methods.

As an example of this kind of research, a brief description is given of how the size of audiences, the potential customers of the advertisers, are measured in the case of commercial television. The following methods are possible.

(a) Telephoning a sample of viewers. They are asked what they are viewing at the time they are called. This method has two main disadvantages: the information obtained is limited to what is being seen at the particular moment of the call, and the sample is not representative, since not every family has a telephone.

(b) Personal interview. The respondent enumerates the programmes he has seen in a given period and his reaction to them. He is also asked what other members of the family saw the programmes and their reactions. (The BBC use this method.)

(c) The "viewing" diary. Selected families keep a diary of the programmes they view for, say, a week and answer a number of questions on them. This method (one used in the USA) will give biased results unless the diary is kept in a reliable manner.

(d) Invited audiences. Representative audiences are invited to to view several programmes, after which they are questioned on their reactions to each programme. This method is also used in the USA.

(e) Automatic recorders. These devices are attached to the television sets of a very carefully selected sample of viewers. They record when the set is on and to what station it is tuned. Market research organisations who use such devices are the Nielsen Company who use the "recordimeter" and the "audimeter" for recording and measuring the amount of viewing, and Television Audience Measurement Ltd., who use the "tammeter".

Two great difficulties face market-research workers. First, no lists are usually available for the relevant populations. There is no list available, for example, of gardeners, which might be useful in the case of market research on gardening implements. Next, random sampling would entail too high a cost. Hence quota sampling is resorted to, with its attendant bias and a sampling error which is not measurable. However, with properly trained interviewers and care it is possible to get useful results.

PUBLIC OPINION POLLS

The most widely known organisation in this country which carries out polls on public opinion is the British Institute of Public Opinion (B.I.P.O.). It is commonly known as the Gallup Poll, and is one of a number of similar dependent institutes throughout the world, all affiliated to the Association of Public Opinion Institutes.

Both random and quota sampling are used. Random samples are used in individual consitituencies in the case of election polls and are based on the electoral register. Quota sampling is used nationally. The controls for quota sampling are: *(a)* regional, *(b)* rural and urban (in the proportion of 1 to 4), *(c)* size of town, *(d)* sex of respondent, *(e)* age group, *(f)* socio-economic group (four divisions), *(g)* political party of sitting member of the constituency. Some 2,000 people are interviewed.

QUESTIONS

1. What are the characteristics of a good questionnaire to be used by interviewers in an opinion survey?   *Institute of Statisticians.*
2. Draft a short questionnaire for a pilot inquiry into house-holders' preferences in floor polishes.

Suggest the types of alterations which you may have to make to the form for the full-scale inquiry when the results of this pilot survey have been scrutinised.   *Institute of Statisticians.*
3. A large transport organisation has asked you to undertake a survey of the travel habits of the population in its area with a view to the better administration of the transport services.

Draft a plan of action to carry out this survey stating what information will be essential; give also some suggestions for further information which you consider might be helpful. Which method or methods would you use to obtain your data?

CHAPTER 5

Preparation for Tabulation

THE DATA

EDITING THE DATA

Editing the recorded data will often be necessary before proceeding to tabulation. This will, in fact, always be the case when dealing with questionnaires.

The process of editing will consist of:

(a) *checking that the schedules are apparently correctly completed;*
(b) *ensuring that there are no inconsistent entries;*
(c) *entering any figures that have to be calculated* (respondents should never be required to make calculations);
(d) *coding the answers, where necessary.*

ANALYSIS OF DATA

The data can be analysed directly from the forms or record cards. In this case tally-sheets are often useful. Squared paper is an aid to neat and hence accurate work. When counting numerous sorted data, it is usual to make strokes on the working paper in fives arranged as in Fig. 2 since groups of five are very easy to count.

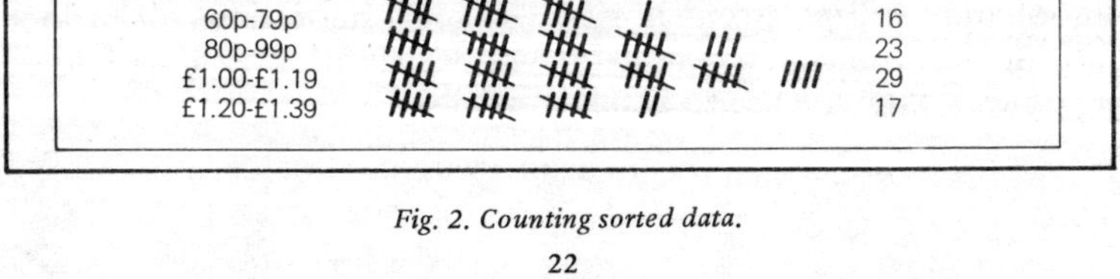

NUMBER OF WAGE-EARNERS AT VARIOUS HOURLY RATES		
		TOTAL
60p-79p	~~IIII~~ ~~IIII~~ ~~IIII~~ I	16
80p-99p	~~IIII~~ ~~IIII~~ ~~IIII~~ ~~IIII~~ III	23
£1.00-£1.19	~~IIII~~ ~~IIII~~ ~~IIII~~ ~~IIII~~ ~~IIII~~ IIII	29
£1.20-£1.39	~~IIII~~ ~~IIII~~ ~~IIII~~ II	17

Fig. 2. Counting sorted data.

22

Often the data are transferred to cards so that the sorting can be done more easily. In this case, coding will be necessary. The cards may be ordinary cards, in which case sorting them into categories will be done manually; with edge-punched cards the sorting is done by means of a needle, and in the case of punched cards mechanically. In some cases, the original record card can also be used as the sorting card. Such cards are known as Dual-purpose cards. An example is shown in Fig. 3.

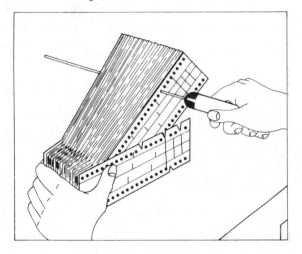

Fig. 3. Sorting edge-punched cards.

CODING THE DATA

CODING

Before the information on the schedules or questionnaires is transferred on to cards for sorting, the information must be coded. Coding consists of giving a number to each classification. Items can then be sorted under code numbers. Examples of coding are as follows.

(a) To designate sex, two code numbers would be required, 1 and 2.

(b) If it were required to classify incomes in four groups, (i) under £800 p.a., (ii) £800 but under £1,400, (iii) £1,400 but under £2,000 p.a., and (iv) £2,000 and over, four code numbers would be required —1, 2, 3 and 4.

Numbers, except when they are to be grouped, do not require to be coded.

EDGE-PUNCHED CARDS

An example is shown in Fig. 3. These cards are made in a number of suitable sizes. They have holes along each edge. On the body of the card can be recorded information if they are to be used as a dual-purpose card. A punch is used to make the round holes into V-shaped notches. They can then be sorted by means of a needle, as shown in Fig. 3. The cards are put in a pack (and to ensure they are the correct way round one corner is cut off), a needle is inserted in the requisite hole and the pack is then lifted and very thoroughly shaken. The cards having the notch fall out. These are the cards recording the information punched.

Since the cards have only a limited number of hole positions, it is necessary to make the utmost use of them. Where it is necessary to make provision for numbers from 1 to 9 to be recorded, four holes can be used, as follows. Let the holes denote values 1, 2, 4 and 7. The other numbers can be coded by using two holes, e.g. to denote 6, holes 2 and 4 would be punched.

These cards have one advantage over other cards, even over those used in the more elaborate systems. It is possible to obtain any particular card more quickly than by any other method, since no elaborate sorter is required.

QUESTIONS

1. Describe the design and uses of any system of small hand-punched cards.

Discuss the relative advantages of different methods of coding numerical information on such cards.

2. Give a short account of the way in which questionnaires are edited.

3. The following is an extract from one of a large number of reports on road accidents collected in 1975. You have been asked to prepare a statistical report on road accidents, and intend to make the analysis with the help of edge-punched cards, on which there are only 40 holes. Design a master card for coding as much relevant information as possible.

Time of day	10.15 a.m.
Vehicles involved	Bus and bicycle
Number of persons injured	1 dead, 2 seriously injured, 1 slightly injured
Weather	Rain
Distance from railway station	2 miles
In speed limit area	No

Class of road	A
State of road surface	Bad
Total number of persons involved	43
Number of witnesses	5
Standard of street lighting	Good

Institute of Statisticians.

CHAPTER 6

Tabulation

Data, however obtained, must be tabulated or put into tables before being used for analysis. Professor Bowley refers to tabulation as "the intermediate process between the accumulation of data, in whatever form they are obtained, and the final reasoned account of the results shown by the statistics."

It is quite obvious that the layout of tables is dependent on the information to be presented and the purpose for which they are prepared. Nevertheless, there are a number of principles which underlie the preparation of all tables. These will be considered in this chapter.

TYPES OF TABLES

INFORMATIVE OR CLASSIFYING TABLES

These are original tables which contain systematically arranged data compiled for record and further use, without any intention of presenting comparisons, relationships or the significance of the figures. In other words, they merely provide a convenient means of compiling data in a form for easy reference, very frequently in chronological order. This type of table is frequently referred to as a schedule.

GENERAL OR REFERENCE TABLES

These tables contain a great deal of summarised information. They are not used for analytical purposes and usually do not give averages, ratios or other computed measures. If they are embodied in a report, they are usually relegated to an appendix. They are often the source from which summary tables are compiled.

TEXT OR SUMMARY TABLES

These tables are used to analyse or to assist in the analysis of classified data. They show only the relevant data of the question being
discussed. Ratios, percentages, averages and other computed
measures are often added. If included in a report, they are found in
the body of the text. These tables are interpretative or derivative
tables, in that they are analytical and are prepared to present significant aspects of the data.

SIMPLE AND COMPLEX TABLES

A simple table presents the number or measurement of a single set
of items having the characteristics stated at the head of a column or
row which forms the basis of the table, as in Fig. 4.

*Food Expenditure of Families in
Receipt of Wages*

Amount per head for food	Number of families
£6.00–£6.99	58
£7.00–£7.99	96
£8.00–£8.99	82
£9.00–£10.00	28
	264

Fig. 4. An example of a simple table.

A complex table presents the number or measurements of more
than one group of items set out in additional columns or rows, and
the table is often divided into sections. Such tables generally show
the relationship of one set of data to another or others, and are
often arranged so that comparisons may be made between related
facts.

In the case of business statistics it is more usual to find complex
tables, because to facilitate a proper consideration of all related facts
full information has to be included. Comparative figures, whether
absolute or percentages or averages of various kinds, are frequently
required, and are therefore incorporated in many forms of tabulated
statistics. In other fields of inquiry, more simple tables are usual.

FREQUENCY DISTRIBUTION TABLES

These give the number of items of different sizes. The example given of a simple table in Fig. 4 was a frequency distribution table.

TIME-SERIES TABLES

These show the values of a variable over a period of time (*see* Fig. 5).

XYZ, Ltd. : Sales

	£
1975	16,237
1976	17,362
1977	23,117
1978	33,206
1979	35,207

Fig. 5. An example of a time series table.

CONSTRUCTION OF TABLES

The first step is to prepare a rough draft. In making a suitable layout, it is usually found necessary to alter the original design. The alteration often consists in changing the rows to columns or the other way round. The cardinal principles involved are clarity and ease of reference. A table should fill the space allotted to it. Equal margins should be left on both sides and more space allowed at the bottom than at the top. The headings of the column and rows should be clear, but concise. A title must be given to the table, which must be clear and bold. Although conciseness is necessary, the title must nevertheless give adequate information. If more than one line is required, the pyramid formation is usual. Sometimes the "pyramid" is inverted thus:

EXPORT OF MACHINERY
UNITED KINGDOM
1960–1970

ARRANGEMENT OF DATA

Data are arranged according to the purpose of the table. This may be according to size or importance, chronologically, geographically, or alphabetically. It may be according to some basis of classification, or merely some usual order. Data to be compared are usually placed in parallel columns, probably adjacent.

Employees in XYZ, Ltd.

Age	Operatives			Administration			Total		
	Male	Female	Total	Male	Female	Total	Male	Female	Total
Under 18									
18 and over									
Total									

Fig. 6. A table in blank for presenting the distribution of employees according to (a) age, (b) sex, (c) employment.

(a) Treatment of units. These must be given fully and accurately, e.g. price per tonne. When a number of columns have the same unit, this is placed over the several columns and not repeated over each.

(b) Spacing and ruling. Clarity, and emphasis, where necessary, can be helped by the judicious use of variations in type, spacing and ruling. Where there are many rows of figures, a break should be made, say, after every fifth row. In the case of, say, monthly figures, a break after every three months.

Printed and typewritten tables can often be set up without the use of ruling, by proper spacing. For instance, major vertical divisions can be separated by a wide space, and related columns can be set closer together.

FOOTNOTES TO TABLES

These are used for four main purposes.

(a) To point out any change in the basis of arriving at the data. For example, sales may be recorded as "ex-factory" for some of the entries and at delivered prices for others. Any heterogeneity in data recorded must be disclosed to avoid wrong conclusions being drawn.

(b) Any special circumstances affecting the data. For example, a fire in the works must be noted.

(c) To clarify anything in the table.

(d) To give the source. This must always be done in the case of secondary data.

QUESTIONS

1. Construct a blank table in which could be shown the number of fires, both serious and slight, in an industrial town. The table should include:

(a) the place of the fire, classified into the groups—factories, warehouses, private dwellings, offices, truck and motor lorries, etc.;

(b) the cause of the fire, classified into carelessness, electrical defects and unknown;

(c) by whom the fire was discovered and extinguished.

2. Lamps are rejected at several manufacturing stages for different faults. 12,000 glass bulbs are supplied to make 40-watt, 60-watt and 100-watt lamps in the ratio of 1 : 2 : 3. At stage 1, 10% of the 40-watt, 4% of the 60-watt, and 5% of the 100-watt bulbs are broken. At stage 2 about 1% of the remainder of the lamps have broken filaments. At stage 3, 100 100-watt lamps have badly soldered caps and half as many have crooked caps; twice as many 40-watt and 60-watt lamps have these faults. At stage 4 about 3% are rejected for bad type-marking, and 1 in every 100 are broken in the packing which follows.

Arrange this information in concise tabular form. Which type of lamp shows the greatest wastage during manufacture?

Institute of Statisticians.

3. Design a blank (or skeleton) table to bring out the differences in direct and indirect taxation in the UK, USA and France for the two years 1970 and 1980.

Take income tax as direct and other taxes as indirect taxation; subdivide the latter if you think it desirable. Both actual and "per capita" figures should be allowed for. Suggest a suitable title.

4. Construct a blank table in which could be shown, at two different dates and in five industries, the average wages of the four groups, males and females, eighteen years and over and under eighteen years.

5. Draft a blank table to present the information below relating to the motor-car industry in the UK for the year 1970 and the four years 1977–1980.

Number of workpeople employed.

Average wage earned.

Exports of motor-cars (number and value).

Imports of motor-cars (number and value).

Number of new registrations.

6. Draft a blank table to show the following information for the United Kingdom to cover the years 1940, 1955, 1970 and 1980.

(a) Population.

(b) Income tax collected.

(c) Tobacco duties collected.

(d) Spirits and beer duties collected.

(e) Other taxation.

Arrange for suitable columns to show also the "per capita" figures for *(b)*, *(c)*, *(d)* and *(e)*. Suggest a suitable title.

7. In a manufacturing firm there are 1,300 persons involved in production, 150 in the sales department, and 100 in administration (these three departments make up the whole of the organisation). 900 men and 450 women are employed. The number of male juveniles is the same as the number of female juveniles—these numbers are in addition to the number of men and women referred to above.

850 men are employed in production and 43 female juveniles. 90 women are employed in the sales department and 16 male juveniles. 29 women are employed in administration together with 36 female juveniles.

Draw up a table to show the complete distribution of men, women, male and female juveniles in the three departments of the firm, together with appropriate totals and sub-totals.

CHAPTER 7

Computation, Accuracy and Approximation

COMPARISONS

When comparing averages, ratios or percentages, it is very important to be sure that the same things are being compared. If, for example, the average wage of two groups of workers are compared, do the two groups of workers contain the same proportion of men and women? Again, if the rate of profit for a number of years is the subject of comparison, is the composition of the goods sold the same? A change in an average, ratio or percentage may be due to an actual change in the particular thing being compared, or it may be due to a change in the constitution of the groups. When investigating such changes, therefore, it will be necessary to consider to what extent the material is heterogeneous. A good example where care is required is the case of crude death rates. It is often found that the death rate of healthy towns is higher than in slums. The reason is, of course, that in slums the people are young, whereas in certain very healthy towns such as Worthing the average age is very high.

Another common error is to average a number of averages and to expect that to be the average of the whole. This can easily be shown to give a wrong result.

Example 1.

	Wages	*Employees*	*Average wage*
Dept. A	£4,000	400	£10
Dept. B	3,200	800	4
Total	£7.200	1,200	£6

£6, which is, of course, the correct average is not the average of £10 and £4.

MIXING OF NON-COMPARABLE RECORDS

Suppose the incomes for a certain airline in respect of a certain route, and the gross profits for two consecutive years are as follows.

Example 2.

	Year 1 £	Year 2 £
Income	140,000	100,000
Gross profit	42,000	24,000
% Profit	30%	24%

Can we say that the rate of profit has fallen?

Consider the following additional analysis:

Example 3.

	Year 1		Year 2	
	Gross profit £	Income £	Gross profit £	Income £
Freight	2,000	20,000	4,000	40,000
Passenger fares	40,000	120,000	20,000	60,000
	42,000	140,000	24,000	100,000

It will be seen that the gross profit on freight is the same for both years, viz. 10 per cent; also the gross profit on passenger fares is the same for both years, viz. 33⅓ per cent. The ratio of profit to income has altered NOT because of a change in rates of profit but because there is a change in the distribution of sources of income bearing different rates of profit. *No comparison is valid which does not allow for differentiation of sources of income.*

FALLACIES ABOUT PERCENTAGES

As an example of an illogical conclusion about percentages, many people would accept the conclusion that because the percentage of deaths to total injuries for aircraft A is 6.7 per cent, and for aircraft B it is only 2 per cent, aircraft B is the safer machine. However, given the following information, it will be seen that this is not so.

Example 4.

	Miles travelled	*Total injuries*	*Injuries involving death*
Aircraft A	10,000,000	1,500	100
Aircraft B	5,000,000	3,000	60

Aircraft A has only 1 death per 100,000 miles whereas aircraft B has 1 death for every 83,333 miles.

STANDARDISATION

An alternative method to sub-dividing the data until the required degree of uniformity is obtained in order to make valid comparisons, is a process known as standardisation. This gives a single rate or average, instead of rates and averages for each sub-division. The standardised rates are weighted averages of the individual rates, weighted according to a standard distribution. In the case of a standardised death rate, the individual death-rates for each age-group would be weighted by a standard age distribution. In Example 1, when comparing wages over, say, a period of years, so that the average only showed changes in wages, the average wage for each department might be weighted in the ratio of 4 to 8, to give a single average. If the averages of the totals were compared, this would not only show changes in wages, but also show changes in the relative numbers in each department.

HAND ELECTRONIC CALCULATORS

Logarithmic tables, tables of squares, interest tables, annuity tables and many others, no matter how comprehensive they are, do not provide entries for all purposes and they are often bulky and unwieldy. Electronic calculators do not have these disadvantages and have thus rendered many tables obsolete.

Slide rules—the cost of a really good one is more than that of the least expensive electronic calculators—are only capable of giving approximate results and depend upon the user having excellent eyesight. They have been relegated to the rank of bygones even quicker than tables.

The great advantages of electronic calculators are:

(a) they cut out boring calculations which can lead to errors;

(b) they are extremely fast.

There are some intriguing things about electronic calculators: the short-cut methods which are sometimes used when calculations are made without the use of such an instrument are not the best methods when using one; nor are the usual formulas always the most efficient ones to use with electronic calculators.

Electronic calculators may be divided into three groups.

Non-programmable calculators. There is a wide range from those that have only the arithmetical functions (+ − × ÷), a single memory

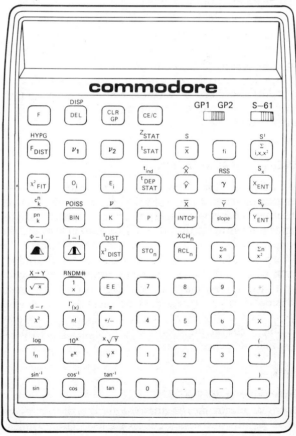

Fig. 7. The "Statistician" keyboard layout.

and, perhaps, a % key to those having a large number of mathematical functions and a number of memories.

Programmable calculators. Programmability has considerable advantages. With a non-programmable calculator, every step in a calculation demands at least one key-stroke. With a programmable calculator, constants and operations can be stored in the right sequence in the calculator ready to operate on the variables as they are entered. The task of the operator is reduced to entering the appropriate variables at the appropriate points. Programmes may be taken from a programme library or devised by the operator. Either way, they are entered simply by keying in a sequence of key-strokes appropriate for the calculation.

Specialised calculators. These are pre-programmed for specific kinds of uses, e.g. navigation, finance. Figure 7 shows the keyboard layout of a calculator—the Commodore S61—for statistics.

There is a key marked χ^2 $_{DIST}$ (five keys down, three across). This key takes the place of a whole complicated programme that would have to be entered in a programmable calculator. This is true also of many other keys.

THE USES OF APPROXIMATION

In many cases, only approximate data are possible. This is so in the case of measurements, for no physical measurement can be exact, although usually the required degree of precision is possible. Exact enumerations are often possible, but again, when they are the result of sampling, they are estimates and approximate. However, providing the sampling is random, the degree of accuracy is measurable.

But even when and if exact data are obtainable, it is often only necessary to use approximate figures. There is no point in spending time and money to get more accurate figures than are necessary for the purpose for which they are required. Often indeed approximate figures enable a clearer picture to be obtained, e.g. a firm's Balance Sheet in which the pence are omitted.

ROUNDING

When approximate values are given by:

(a) *stating the value to the nearest unit,* e.g. the nearest tonne or the nearest £, or to the nearest multiple of some unit, e.g. to the nearest 1,000 tonnes;

(b) *giving the next highest or next lowest whole number,* e.g. £3.29 given as £3 or £4;

(c) expressing the value to so many significant figures, e.g. 3.475 expressed correct to 2 significant figures as 3.5; the figures are said to be rounded.

Note the following apparent error due to rounding:

Example 5

Actual	Rounded
3,876	4,000
1,777	2,000
514	1,000
6,167	6,000

The addition of the rounded figures appears to be wrong, but the addition must not be made formally correct, otherwise the total would not be correct to the nearest 1,000.

STATEMENT OF DEGREE OF ACCURACY

The extent of the error must be stated. This can be done in a number of ways, e.g.:

(a) 3,000 tonnes to the nearest thousand tonnes,
(b) 3,000 ± 500 tonnes,
(c) 3.35 correct to three significant figures,
(d) 562 to the nearest whole number,
(e) 56 ± 3 per cent.

KINDS OF ERROR

The error which arises when an approximate or estimated value is used instead of the actual value must not be confused with errors which arise because of mistakes made in counting, measuring, calculating or in making observations, which are best called mistakes. Error refers to differences between the actual value and the estimated or rounded value.

Such errors may be due to estimating by means of a sample—the smaller the sample, the greater the error; they may arise because of bias or because the data have been recorded approximately.

RELATIVE AND ABSOLUTE ERRORS

The absolute error is the actual difference between the actual value and the estimated or rounded value. The relative error is the ratio, often expressed as a percentage, of the absolute error to the actual figure or where, as is often the case, this is not known, to the estimated value. For example, if output is 40,000 tonnes to the nearest 1,000 tonnes, then the maximum absolute error is 500 tonnes and the maximum relative error is 500 divided by 40,000, i.e. 0.0125 or 1.25 per cent.

LAWS OF ERRORS

(a) *The absolute error of a sum equals the sum of the absolute errors of its components.*
Add 500 (to the nearest 10) and 400 (to the nearest 100).
$$(500 \pm 5) + (400 \pm 50) = 900 \pm 55.$$

(b) *The absolute error of a difference equals the algebraic difference of the errors of its components.*
From 500 (to the nearest 10) subtract 400 (to the nearest 100).
$$(500 \pm 5) - (400 \pm 50) = 100 \pm 55.$$

(c) *The relative error of a product is equal to the sum of the relative errors of its components.*
Multiply 500 (to the nearest 10) by 400 (to the nearest 10).
$$(500 \pm 1\%) \times (400 \pm 1\frac{1}{4}\%) = 200,000 \pm 2\frac{1}{4}\%.$$

(d) *The relative error of a quotient is equal to the algebraic difference of the relative errors of its components.*
Divide 500 (to the nearest 10) by 400 (to the nearest 10).
$$(500 \pm 1\%) \div (400 \pm 1\frac{1}{4}\%) = 1.25 \pm 2\frac{1}{4}\%.$$

The last two laws are only approximately true, and then only provided that the degree of error is small.

Note that in the example given under rule (d) it is the possible error that must be calculated, i.e. the greatest error possible. This will be when 500 + 1 per cent is divided by 400 − 1¼ per cent; the relative error is then 1 per cent − (−1¼ per cent), i.e. 2¼ per cent. Similarly, the relative error of the other limit is −1 per cent − (+1¼ per cent), i.e. −2¼ per cent.

BIASED ERROR

When the error is all in one direction, it is said to be biased or cumulative. Thus amounts which are given to the next highest whole number are biased.

COMPENSATING ERRORS

When the errors are such that they tend to cancel each other, they are said to be unbiased or compensating. Thus, numbers approximated to the nearest whole number are likely to give rise to compensating errors.

ACCURACY AND ERRORS

When errors are biased, the absolute error of a sum will be greater than the errors of the individual items—the greater the number of items, the greater the error. In the case of subtraction, the absolute error may be reduced. In the case of multiplication, the relative error will be increased, but in the case of division, the relative error will be decreased.

Where the number of quantities is large and the errors are compensating, the resultant error of their sum or difference is reduced.

QUESTIONS

1. Explain the meaning of the statistical terms:

(a) compensating error, and

(b) biased error.

The numbers 56,000, 7,000 and 20,000 are correct to 5 per cent, 0.5 per cent and 0.05 per cent respectively. Calculate the absolute error in the aggregate of the three numbers, Express this absolute error as a percentage of the aggregate value.

2. Explain "statistical error." Approximate the following figures to:

(a) Nearest thousand £.

(b) Nearest hundred £.

(c) Next thousand £ below.

Calculate the absolute and relative errors of their totals.

Exports of Mining Machinery from the United Kingdom, May 1972

To West Africa	34,053
" South Africa	12,865
" India	33,233
" Malaya	8,177
" Australia	45,847
" Other Commonwealth countries and Irish Republic	67,581
" Poland	12,461
" France	18,346
" Other foreign countries	78,455

3. Give specimen illustrations of the meaning of the term "error" in statistical work and, hence, show how this differs from the ordinary interpretation of the word.

Chartered Institute of Transport.

4. Define statistical error. Find the values of:

    $(19 \pm 2) + (85 \pm 6)$
    $(76 \pm 5) - (15 \pm 1)$
    $(40 \pm 7) \times (25 \pm 4)$
    $(480 \pm 20) \div (120 \pm 5)$

5. In statistical work it is often necessary to use approximate figures. Show how the degree of accuracy in the result can be determined in the course of arithmetical manipulation of such data.

6. What do you understand by the estimatation of errors? The total profits of 144 companies amount to £7,425,000. Estimate the possible error when each company's profits have been approximated to:

    (i) the nearest thousand £;
    (ii) the next thousand £ above the actual figure.

7. Rewrite the following table giving the figures correct to the nearest 1,000. What is the amount of:

    (a) the absolute error of the total, and
    (b) its relative error?

Suppose *your* figures were the only ones available; then state:

    (c) the percentage of emigrants who went to Canada, and
    (d) the amount of relative error in the above percentage, and show its relation to the relative error of the number who went to Canada and the relative error of the total.

Emigrants of British Nationality Travelling Direct by Sea from the United Kingdom to Extra-European Countries, 1950

By Destination

Canada	13,434
Australia	54,184
New Zealand	10,562
South Africa	9,320
Other British countries . . .	25,434
Foreign countries	17,304
	130,238

Source: *External Migration, HMSO.*
Institute of Statisticians.

8. A Company has 2 factories. From the data below compare the accident rates in the two factories.

Ages	Factory X		Factory Y	
	Employees	Accidents	Employees	Accidents
Under 21	200	28	650	91
21–30	250	25	350	35
30–40	500	40	150	12
40 and over	50	2	50	2
	1,000	95	1,200	140

CHAPTER 8

Design of Forms

Much of the information that is subsequently summarised and presented in the form of tables is first recorded on a form. One type

TYPE OF ENTRY SPACES

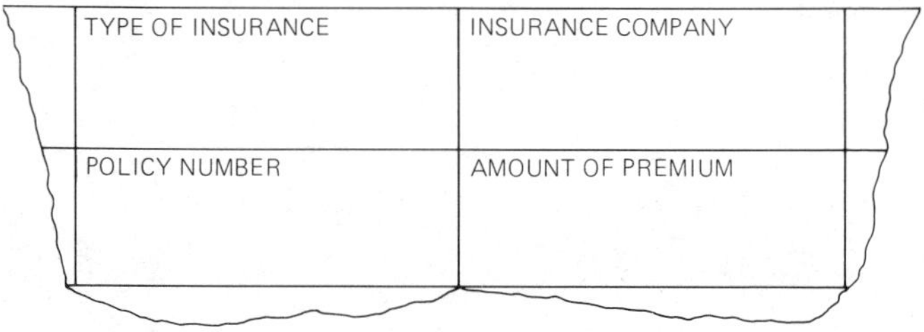

| TYPE OF INSURANCE | INSURANCE COMPANY |
| POLICY NUMBER | AMOUNT OF PREMIUM |

Fig. 8. Panel arrangement.

of form is the questionnaire. But in any case, whatever the type of form, there are certain principles to be observed in its design.

THE PURPOSE OF FORMS

In a Stationery Office publication, *The Design of Forms,* it is stated: "Forms are used to record and communicate information in order that the information shall be set out in an orderly, pre-determined way and a regular routine established for entering and using the information and handling the form itself."

HOW TO DESIGN A FORM

Before the actual designing of the layout, careful consideration must be given to the following matters, all of which affect the design if the form is to fulfil its proper functions.

(*a*) The information wanted.

(*b*) How will the form be used? Who will enter the information, method of sorting, tabulating, filing, where the form is to be used, e.g. in the factory, outdoors, the frequency of use.

In the actual designing of the layout, careful attention should be paid to the following points.

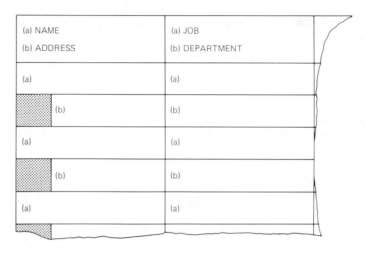

Fig. 9. Columnar entries.

Notes: (a) Each column is used for two entries, thus avoiding too wide a form.
(b) The shading in the first column makes the name stand out more easily in a long list.

MARK SHEET													
NAME OF STUDENT	MATHS	ENGLISH	HISTORY	GEOGRAPHY	CHEMISTRY	PHYSICS	ECONOMICS	FRENCH	MUSIC	P.T.	DRAWING	SCRIPTURE	LA

Fig. 10. Headings for narrow columns.

(a) The form should not look complicated.
(b) The captions should be clear and precise.
(c) Instructions (which should be adequate) should, if lengthy, be given in a part of the form distinct from the questions.
(d) Give the form some identification (e.g. a title).

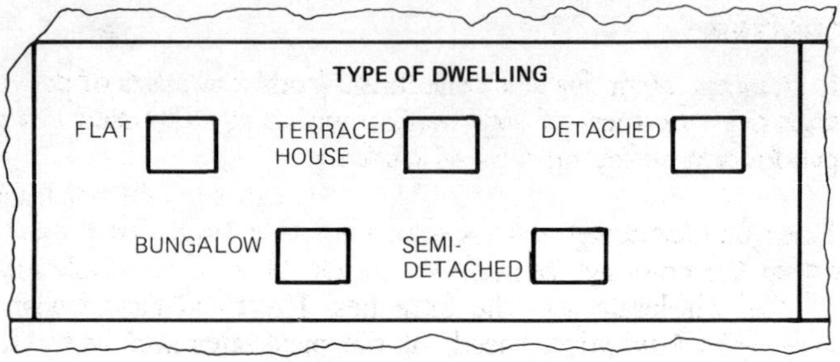

TYPE OF DWELLING

FLAT ☐   TERRACED HOUSE ☐   DETACHED ☐

BUNGALOW ☐   SEMI-DETACHED ☐

Fig. 11. "Boxes": the appropriate box will be ticked.

Details	Last month		This month		Increase or Decrease	
	£	%	£	%	£	%
Gross sales						
Less returns and allowances						
Net sales						
Less prime cost of sales						
Gross profit on sales						
Less selling expenses						
Less general expenses						
Net profit on sales						
Add other income						
Net income						
Less fixed charges						
Add balance brought forward						
Amount available						
Appropriations						
To carry forward						

Fig. 12. An example of a form used by accountants.

(e) Arrange the entries so that the form is easy to read and it is easy to make the entries.

(f) Spacing must receive special care. Remember this is determined by the entry to be made, not by the length of the caption.

QUESTIONS

1. Design a form for the collection, from a number of provincial distributive branches, of information which is to be used as a guide to production policy on sales of clothing.

Institute of Statisticians.

2. A manufacturing organisation has selling branches in each large town in the country. It makes 6 kinds of articles which are sold retail and wholesale by the branches. The Head Office wishes to plan a Sales Campaign based on the past sales and likely future demand. Design a form for the collection of the necessary data and draft the instructions for completing the form.

Institute of Statisticians.

3. Design a form for recording the daily output from a set of six similar machines, with spaces for entering monthly totals and appropriate comparisons of performance of the different machines.

Institute of Statisticians.

CHAPTER 9

Charts and Diagrams

Statistical data can often be presented in chart form or by means of diagrams or graphs, in addition to being presented in tabular form. Often this enables relationships and trends and comparisons to be grasped more readily.

TYPES OF CHART

BAR CHARTS

These charts enable magnitudes to be compared visually. Bars are drawn whose length is proportional to the magnitude to be represented. Thus, if it were required to compare a sales figure of £2,000 with a sales figure of £2,500, two bars would be drawn whose lengths were proportional to these two amounts, for example, 2 centimetres and 2½ centimetres respectively.

When data are to be presented in the form of bar charts, a suitable scale must be chosen, and this will be indicated either at the side or the bottom of the diagram. This scale MUST start at zero, otherwise false impressions are given. The choice of horizontal or vertical scales is optional when items in a group are to be compared, as for example, a bar chart comparing the populations of the continents. When, however, the data are to be charted with reference to a series of dates, the bars should be drawn vertically, so that the dates appear horizontally, and the scale is shown vertically (*see* Fig. 13).

The bars are usually separated, the width is entirely a matter of neatness of presentation. A chart, like a table, must have a title.

COMPONENT BAR CHARTS

These charts show visually the way in which a "whole" is divided.

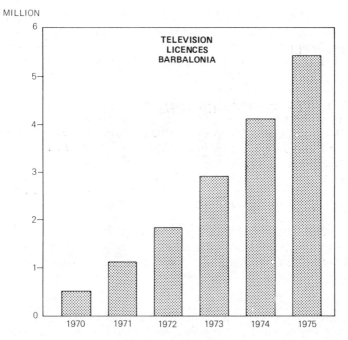

Fig. 13. A simple bar chart.

Thus a bar chart, representing sales, might be divided into two parts, representing home sales and export sales (*see* Fig. 14: note the key, and the method of shading). Of course, the divisions would be proportional to the respective sales.

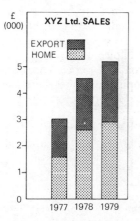

XYZ Ltd.—Sales (£000)			
	1977	1978	1979
Export	1.4	1.8	2.29
Home	1.6	2.7	2.9
Total	3.0	4.5	5.18

Fig. 14. A component bar chart.

THE PERCENTAGE COMPONENT BAR CHART

In this case, the bars are divided in proportion to the percentages that the parts bear to the whole. The scale will be a percentage scale, and all charts will be the same length. A key will also be necessary (Fig. 15).

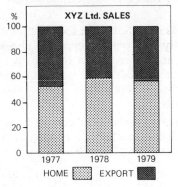

Fig.15 A percentage component bar chart

THE MULTIPLE BAR CHART

This is sometimes known as a compound bar chart. This chart groups two or more bar charts together. More than one set of comparisons can be made. Figure 16 gives an example of such a chart. Here exports and imports can be compared from year to year. In addition, imports can be compared from year to year. Finally, exports can also be compared over the years.

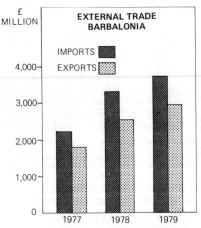

Fig. 16. A multiple bar chart.

PICTOGRAMS

These are really a form of pictorial bar chart, a useful way of presenting data to people who cannot understand orthodox charts. Small symbols or simplified pictures represent the data. There are important rules to follow if pictorial charts are to fulfil their function.

(a) The symbols must be simple and clear.

(b) The quantity each symbol represents should be given.

(c) Larger quantities are shown by a great number of symbols, and not by larger symbols. This is a frequent error. A part of a symbol can be used to represent a quantity smaller than the value of a whole symbol.

Pictograms are used to make visual comparisons in exactly the same way as a bar chart, and, in effect, the number of symbols corresponds to the length of a bar chart (Fig. 17).

PEGASUS VAN MANUFACTURING CO., LTD.
OUTPUT OF VANS

				= 1,000 VANS

OUTPUT
1976 2,004
1977 2,996
1978 4,219
1979 5,324

Fig. 17. A pictogram.

PIE CHARTS

Like the component bar charts, pie charts show the relationship of parts to the whole. However, there is one important difference. In the case of component bar charts, the length of bars are compared, whereas in the case of pie charts, areas of segments are compared. It is, however, difficult to compare areas visually. For this reason pie charts are an inferior form of presentation. They are extremely popular, except among statisticians. Figure 18 is an example of a pie chart. Often the percentages are inserted in the segments, so that, in effect, it is the figures that are compared.

Construction of a pie chart. A circle consists of 360°. The proportion that each part bears to the whole will be the corresponding proportion of 360°, which will require to be calculated. In the example given, taxation is one-fourth of the whole. The segment relating to taxation will therefore have an angle of one-fourth of 360°, i.e. 90° at the centre.

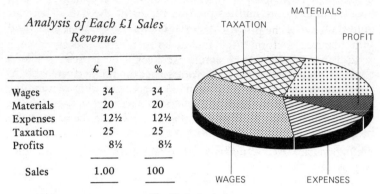

Analysis of Each £1 Sales Revenue

	£ p	%
Wages	34	34
Materials	20	20
Expenses	12½	12½
Taxation	25	25
Profits	8½	8½
Sales	1.00	100

Fig. 18. A pie-chart.

BREAK-EVEN CHARTS

A break-even chart shows the profit or loss for any given output. It also shows the point at which costs and revenue are equal. This is the break-even point. The simplest form of chart shows two curves; one showing the relationship between revenue and output, the other the relationship between cost and output. In the example given, the break-even point is shown as 3,500. At this output, revenue exactly covers costs. Below this output there is a loss; above, there is a profit. Although the chart given shows straight lines, in practice the lines will probably be curves.

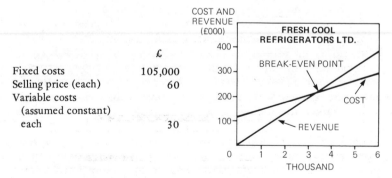

	£
Fixed costs	105,000
Selling price (each)	60
Variable costs (assumed constant) each	30

Fig. 19. A break-even chart.

GANTT PROGRESS CHARTS

There are many kinds of Gantt charts. They are all a particular type of bar chart, and they are all scaled in units of time. The one dealt with is a progress chart (*see* Fig. 20).

In the case of the progress chart, actual performance is compared with planned performance. As will be seen from the chart, each period—in the example shown, a week—will be represented by a certain length. Each length represents the planned performance. As these are not necessarily equal, the same length, representing the same length of time, does not represent the same quantities. The actual performance is charted, and in addition, the cumulative performance is shown. The weekly results are shown by a thin line, the cumulative by a thick line (*see* Fig. 20).

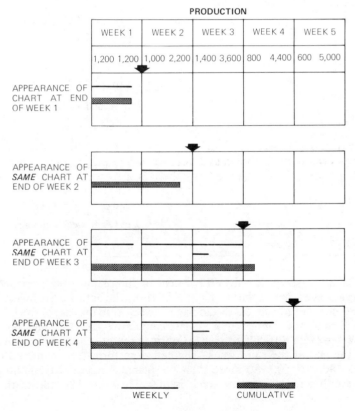

Fig. 20. A Gantt progress chart.

Output
Weekly results — Thin line

Week	Quota	Actual	Actual as % of quota
1	1,200	960	80
2	1,000	1,000	100
3	1,400	1,820	130
4	800	540	67.5
5	600	—	

Cumulative results — Thick line

Week	Quota to date (cumulative)	Actual to date (cumulative)	Actual to date in terms of weeks' quotas
1	1,200	960	960 (i.e. $\frac{960}{1,200} = 80\%$) of first week's quota
2	2,200	1,960	1st week's quota + 760 (i.e. $\frac{760}{1,000} = 76\%$) of second week's quota
3	3,600	3,780	3 weeks' quotas + 180 (i.e. $\frac{180}{800} = 22.5\%$) of fourth week's quota
4	4,400	4,320	3 weeks' quotas + 720 (i.e. $\frac{720}{800} = 90\%$) of fourth week's quota
5	5,000	—	—

In the first week only 80 per cent of the planned performance was achieved; hence the line charted to show the actual performance is 80 per cent of the length allocated to week 1. In the case of week 3, more than one line is required. The cumulative line, the thick one, is a bit more difficult to chart. Consider the position of the chart at the end of the third week. The actual performance at that date is 3,780, that is, 180 more than the planned cumulative amount. The quota for the fourth week is 800. Therefore 180/800 of the fourth week's quota has been done. Therefore the cumulative chart will be carried forward to a point in the space allotted to the fourth week 180/800 of the length allotted to the fourth week. Note the quotas and the cumulative quotas entered at the top of each week's column.

STATISTICAL MAPS

When statistical data refer to geographical areas, often a suitable form of presentation is the statistical map or cartogram.

Fig. 21. Sales according to salesmen's area.

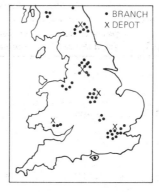

Fig. 22. Distribution of branches.

The main types of such maps are as follows.

(a) The hatched or shaded map. The varying size of the data is denoted by a different shading or hatching—the greater the size, the denser the shading or the closer the hatching. Figure 21 shows such a map. This type of map is also used to show averages and ratios relating to various areas.

(b) Dot maps. These show the number of occurrences. They may be placed at their exact location, as in Figure 22, or by a single dot placed in the area to which they refer, varying in size according to the number they represent, as in Figure 23.

(c) Maps presenting graphs or charts. In many cases, charts or graphs can be placed on the areas to which they refer. Figure 24 is an example of such a map.

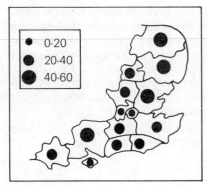

Fig. 23. Number of salesmen in
south-eastern counties.

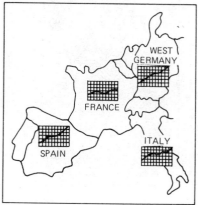

Fig. 24. Growth of sales
to certain European countries.

QUESTIONS

1.   *Exports of Textiles 1979*
          (Woven Piece Goods—Million Square Metres)

	April	May	June	July	Aug	Sept	Oct	Nov	Dec
Cotton	96	78	72	65	77	71	67	73	53
Wool	13	10	9	10	10	9	8	7	6

Make a graphical comparison, by means of a compound bar-chart,
of the volume of exports given in the above table. Use the chart to
write a brief report on the exports of woven piece goods during the
last nine months of 1979.

2. Show the following information on a suitable diagram and give comments upon any features of interest to the transport industry:

Estimated Population in Zones (000)

Zone radius (kilometres)	Manchester		Stockholm	
	1950	1975	1950	1975
1	231	199	28	15
2	415	274	212	190
3	342	319	162	172
4	265	270	87	96
5	80	82	73	97
6	32	36	86	123
7	28	32	72	115
8	3	4	63	135

Chartered Institute of Transport.

3. The following table shows how British Airways earned and spent an average pound (£) of revenue and expenditure during a certain year.

Revenue	p	Expenditure	p
Passenger . .	79	Pay and allowances . .	44
Mail . . .	9	Fuel and oil . . .	16
Freight . . .	8	Aircraft maintenance .	10
Commissions .	2	Commissions . . .	6
Other . . .	2	Passenger and cargo .	
		services	5
		Aircraft standing charges .	4
		Landing fees . . .	4
		Sales and publicity . .	3
		Other	8

Present this information diagrammatically:

(a) using pie diagrams, and

(b) using any other form of diagrammatic presentation of your own choice, giving reasons for your choice.

Chartered Institute of Transport.

4. Describe two forms of percentage distribution diagrams.

5. Describe each of the diagrams listed below and give an illustration in each case:

Bar diagram   Break-even chart
Compound bar diagram   Percentage bar diagram
Progress chart   Pictogram

6. How would you present statistical data referring to geographical areas on maps?

7. The budgeted production and the actual quantities produced of a certain component manufactured by XL Motors Ltd. were as follows:

Month	Budget	Actual
January	1,000	800
February	1,500	1,600
March	2,000	1,800
April	2,000	2,100
May	2,500	2,800
June	3,000	2,900

Show the Gantt progress chart as it would appear at the end of (a) April, (b) June.

CHAPTER 10

Network Planning

Any operation, project or task which is undertaken—however simple or complicated—can be divided into a number of activities. Building a house, for example, could be analysed into a large number of activities, among which would be found, say, drawing up the plans, digging the foundations, ordering bricks and so on. Some activities must be completed before others are started whereas some activities can be carried on at the same time whilst all activities must be performed in an orderly manner so that the project can be completed in the most efficient way.

When the project consists of a large number of activities planning becomes extremely difficult. *Network analysis* provides a technique for efficient planning of any operation. It can also be used very effectively for comparatively small tasks which are usually done by rule-of-thumb methods.

DRAWING A NETWORK

DATA REQUIRED

The following table provides the information required to construct a network.

PROJECT ANSCO

Activity	Immediately preceding activities	Immediately following activities	Duration of activity (days)
A	C	D H	3
B	E	F	7
C	E	A	12
D	A	F	5
E	G	B C	8
F	B D H	None	4
G	None	E	2
H	A	F	9

In respect of *each* activity it is necessary to know the following.

(a) The immediately preceding activity (or activities). As soon as the immediately preceding activity (or activities) are finished it is possible for the activity in question to be started.

(b) The immediately succeeding activity (or activities). It follows that if activity E immediately precedes C then C immediately follows E.

(c) The duration of the activity. This is the time required to complete the activity.

SYMBOLS USED IN DRAWING A NETWORK

An activity is denoted by an arrowed line joining two circles known as *nodes* or *events* which denote the start and finish of the activity. The events are numbered; an activity can thus be identified by the numbers in the nodes at the start and finish of the activity, e.g. 5—6. This is shown in Fig. 25.

Fig. 25. *Symbols used to denote an activity.*

The arrow points towards the immediately following activity; it must never point towards the preceding one.

THE USE OF DUMMY ACTIVITIES IN A NETWORK

A *dummy activity* represented by a broken arrowed line is a nonexistent activity, and thus takes no time to perform and uses no resources.

An activity is identified by the numbering of the nodes at the start and finish of the activity.

Where activities can be carried out simultaneously, that is, each activity has the same preceding activities and the same following activities, then, since each activity must have a different identity reference, a dummy activity is introduced into the network.

In project Ansco, where activities D and H are both preceded by A and both followed by F, a dummy activity is introduced into the network as shown in Fig. 26.

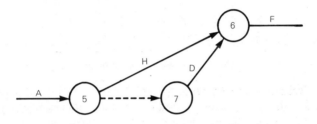

Fig. 26. Dummy activity to ensure activity identity.

Consider an activity A which immediately follows activities X *and* Y and an activity B which immediately follows *only* X. Again it is necessary to introduce a dummy activity—this time to ensure that the network reflects the facts. This will be clear from Fig. 27.

The direction in which the dummy arrow points is very important.

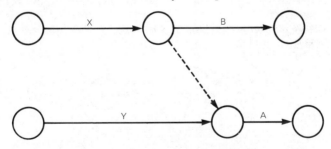

Fig. 27. Dummy activity to preserve logicality.

There is a tendency for students to introduce unnecessary dummies. They must not be introduced other than for the two purposes just described.

LOOPING AND DANGLING

These are errors which must be avoided in drawing the network. *Loops* are illogical. Referring to Fig. 28, event 9 occurs after event 7 and event 8 after event 9; event 8 must therefore occur after event 7 and *not* before as shown in the network.

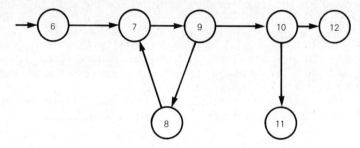

Fig. 28. Looping and dangling in a network.

Dangling arises when an activity such as 10—11 in Fig. 28 is shown with no outcome; nothing follows event 11. All events except the first and last must have at least one activity starting with it and one activity finishing with it.

A TYPICAL NETWORK

Fig. 29 shows the network drawn from the data given. The events have been numbered from left to right and from top to bottom of the network. This is the usual method of numbering.

The activity has been described—in this case by a letter—above the arrowed line and below the line the duration of the activity has been given. *Note that the length of the arrowed lines is* NOT *proportional to the duration of the activity*. The lines are always drawn straight and not curved.

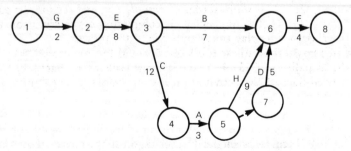

Fig. 29. Network for Ansco project.

THE CRITICAL PATH AND THE DURATION OF PROJECT

Starting from event 1 and calling this zero time, then the earliest time event 2 can take place, that is the earliest time activity E can start is $0 + 2 = 2$ days after event 1 when activity G was started, since the duration of G is 2 days. The earliest time event 3 can take place, that is, the earliest time B and C can start is $2 + 8 = 10$ days after event 1. The earliest time event 6 can take place if only activity B precedes it is $10 + 7 = 17$ days, *but* this is not the case. C, A and H also precede it and all these activities have to be completed before F can start. The earliest time event 6 can take place is therefore $10 + 12 + 3 + 9 = 34$ days after event 1.

Where there are two or more paths to an event the earliest time for the event is the time taken by the longest path.

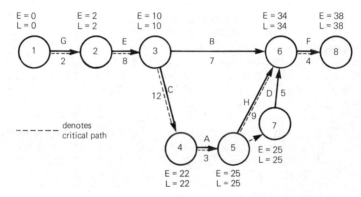

Fig. 30. The critical path, Ansco project.

The earliest time for the final event gives the duration time of the project, that is, the shortest time to complete it, and the longest path is the critical path. Any delay in completing any activity on the critical path will delay the completion of the project.

In Fig. 30 the earliest times are denoted by the letter E and the critical path (1–2–3–4–5–6–8) by a broken line parallel to the arrowed lines of the activities on the critical path. The duration of the project is also shown to be 38 days.

FLOAT

The differences between the duration of the longest path—the critical path—and the alternative paths are slacks or floats; the activities on these alternative paths except those shared with the longest path can be delayed without affecting the duration of the project.

Calculating the amount of float in respect of all those activities not on the critical path enables the planner to know the extent of the excess resources, men and/or materials, and with what activities they are associated.

CALCULATION OF FLOAT

The earliest time to finish the project is also the latest time to finish it. In project Ansco this is 38 days. The latest time for event 6 to occur if the project is to finish on time is $38 - 4 = 34$ days since activity F takes 4 days. The latest time event 3 can take place if only activity B follows it is $34 - 7 = 27$ days. But C, A and H also follow. The latest time for event 3 is therefore $34 - 9 - 3 - 12 = 10$ days. *Where there are alternative paths from an event, the latest time for an event is the lowest figure.* It will be noted that the latest times for an event to take place, that is for the following activity or activities to start, is denoted by L on Fig. 30 and Fig. 31. It will also be noted that for all the events on the critical path the earliest and latest times are identical; the activities must start and end at the times indicated for the project not to be delayed.

For the activities not on the critical path there is a float. Those activities do not necessarily have to start at the earliest time nor finish at the latest time; for example, activity B has 24 days between the earliest time it can start and the latest time it can finish and it only needs 7 days. It therefore has $24 - 7 = 17$ days float. This is the *total float*. It means that the duration of activity B can be expanded by 17 days.

Consider an activity which is not on a critical path such as is depicted in Fig. 31. *Total float* = Latest time to end activity *less* Earliest time to start activity *less* Duration of activity. In the case of the activity depicted in Fig. 31 this is $40 - 8 - 15 = 17$.

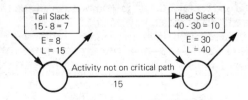

Fig. 31. An acitivty with float.

This is the total amount of time by which the activity could be expanded without affecting the duration of the project, *but* only if the activity started at the earliest time and ended at the latest time. This condition could well affect the floats of the other activities, that is the time by which they could be expanded.

Free float is the amount the activity can expand without affecting the float of the following activity. *Free float* = Total float *less* Head slack (that is the latest time an activity can finish *less* the earliest). In this case, the head slack is $40 - 30 = 10$ and the free float is therefore $17 - 10 = 7$.

Independent float is the amount an activity can expand without affecting the floats of *both* the previous and following activities. *Independent float* = Free float *less* Tail slack (that is the latest time an activity can start *less* the earliest). In this case, the tail slack is $15 - 8 = 7$ and the independent float is therefore $7 - 7 = 0$.

Listed below are the activities of project Ansco which are *not* on the critical path together with their earliest and latest starting and finishing times and also their duration (*see* Fig. 30).

Activity	Starting time Earliest	Latest	Finishing time Earliest	Latest	Duration
B	10	10	34	34	7
D	25	25	34	34	5

The *free float* is calculated as follows:

Activity	Total float	Head slack	Free float
B	$34 - 10 - 7 = 17$	$34 - 34 = 0$	$17 - 0 = 17$
D	$34 - 25 - 5 = 4$	$34 - 34 = 0$	$4 - 0 = 4$

SEQUENCED GANTT CHART

A very useful way of presenting the planning and scheduling of a project is by means of a Gantt bar chart where the activities are shown against a time scale. The chart for the project Ansco is shown in Fig. 32.

The activities comprising the critical path constitute the first bar. The other activities will be represented by other bars parallel to the first bar and placed beneath it; each bar representing a branch of the network. The length of the bars are proportional to the duration of the activities *plus* their free float. The relationships between the branches are indicated by vertical broken lines.

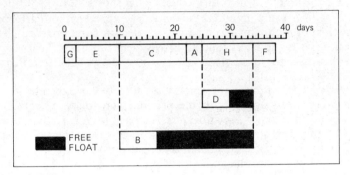

Fig. 32. Sequenced Gantt chart.

QUESTIONS

1.

The "AJAX" project.

Activity	Immediately preceding activities	Immediately succeeding activities	Duration of activity (days)
A	none	B, D	3
B	A	C, G	6
C	B	I	8
D	A	E, F	2
E	D	H	5
F	D	H	4
G	B	H	2
H	E, F, G	none	9
I	C	none	12

You are required to: *(a)* draw the network; *(b)* indicate the critical path; *(c)* draw a sequenced Gantt chart.

2. A company has decided to build new offices on a plot of land which it owns, and a list of the necessary activities has been drawn up by the building contractor. These activities are given in the table below:

NETWORK PLANNING

Activity	Description	Preceding activities	Duration (weeks)
I	Draw up plans	—	3
II	Order furniture and office equipment	I	1
III	Level the site	I	3
IV	Mark out site	III	2
V	Lay drainage	IV and II	4
VI	Make approach roads	IV and II	3
VII	Lay foundations	V	7
VIII	Erect walls	VII	12
IX	Lay paths	VI and VIII	4
X	Erect roof	IX	2
XI	Erect internal walls	X	5
XII	Complete building	XI	10

Set out the network for these activities and identify the critical path. Thus indicate the least time which will be taken to complete the new offices on the basis of the above durations.

Institute of Charted Secretaries and Administrators.

3. Following the receipt of an application for a mortgage loan by an insurance company, it is found that one day is taken in establishing all the facts required to identify the nature of the property and the amount of the applicant's income. Credit references are then taken up, and at the same time surveyors are asked to value the property. It takes 5 days for references to be obtained and 7 days for a survey to be completed. When credit references and a survey are available the application can proceed to the mortgage committee, and it takes 5 days for a decision to be made. Whilst committee consideration is being given the Accounting Dept. is being consulted to determine when funds will be available for the loan, and the rate of interest to be charged. It takes three days for the Accounting Dept. to make its decision. Finally when committee approval has been gained, and it is known that funds are available, the application can proceed to the Board of Directors for sanction. It takes 7 days for this last process to be complete. How long does it take from the receipt of the application form to final agreement by the Board to gain sanction for a loan? Show the details of your estimate in the form of a network diagram and indicate the critical path.

Institute of Chartered Secretaries and Administrators.

CHAPTER 11

Graphs

The movement of data can be presented very effectively by means of a graph, sometimes known as a line chart.

THE PURPOSE OF A GRAPH

A graph shows, by means of a curve or straight line, the relationship between two variables; for example, the amount of sales and the period of time when the sales were made, or the relationship between output and cost. One of the variables will be the independent, and the other the dependent, variable. The formulation of the problem will determine which variable is chosen as dependent. For example, the relationship between the number of tonnes of wheat per hectare and the amount of rainfall can be formulated in two ways. Given the rainfall, what is the output per hectare? or, given the output per hectare, what is the rainfall? The first would be the more usual way. The output per hectare would depend on the rainfall. The output would be the dependent variable, the rainfall, the independent variable.

HISTORIGRAMS

When one of the variables changes over time, e.g. population, or sales, i.e. when the values of the variables form a time series, the graph which shows this relationship is known as a historigram (*see* Figs. 33 and 40).

Note. Do not confuse a historigram with a histogram, an entirely different type of diagram, illustrated in Fig. 47.

Daily Output	
Day	Output
1	100
2	300
3	600
4	500

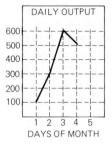

Fig. 33. A historigram.

CONSTRUCTION OF GRAPHS

The first step in the construction of a graph is the drawing of two lines at right angles. These lines are known as co-ordinate axes (*see* Fig. 34).

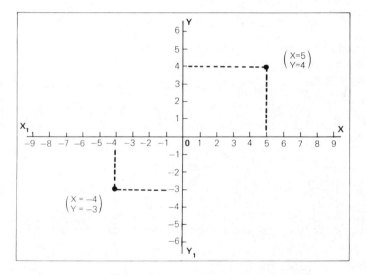

Fig. 34. The construction of graphs.

The axis YOY_1 is called the ordinate or, less mathematically, the vertical axis; XOX_1 is called the abscissa, or the horizontal axis. Other terms are the y-axis and the x-axis respectively.

Both these axes will have a scale, chosen so that the graph will fit the paper, will include all the observations and will avoid too sharp or too flat a curve.

The point of intersection of the two axes is called the origin, denoted by O in the diagram. It will be observed that the co-ordinates form four quadrants. In business statistics only the quadrant YOX is usually of interest.

Having drawn the axes, and marked a suitable scale along both axes, the next stage is to plot the observations. This is best explained by means of examples. Suppose the value of $y = 4$ when the value of $x = 5$ and the value of $y = -3$ when the value of $x = -4$. These two observations are plotted as follows.

The first point is chosen at a distance of 5 (measuring along the x-axis) from the y-axis and at a distance of 4 (measuring along the y-axis) from the x-axis. Similarly, the second point is plotted at a position 4 units to the left of YOY_1 and 3 units below XOX_1 (see Fig. 34).

RULES FOR MAKING GRAPHS

(a) *Independent variable.* Decide which variable is the independent variable. In the case of a time series, time is the independent variable.

This is the x variable. In the case of a time series, the x-axis will, therefore, be scaled in years, months, etc.

(b) *Zero line.* When the graph is concerned with absolute changes in quantities, the zero line must be shown, otherwise the graph will give a false impression (see Figs. 35 and 36).

(c) *Co-ordinate lines.* The graph should be clearly distinguished from the co-ordinate lines (see Fig. 36).

(d) *100 per cent lines.* When the 100 per cent line is the basis of comparison, this should be emphasised (Fig. 37).

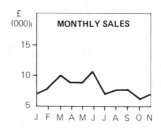

Fig. 35. *Illustration of Rules (a) and (b).*

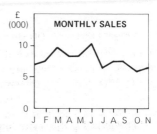

Fig. 36. *Illustration of Rules (b) and (c).*

(e) Clarity. It is better not to show more co-ordinate lines than are necessary to enable the graph to be read.

(f) Lettering. Lettering should be shown horizontally, not vertically.

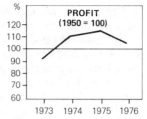

Fig. 37. *Illustration of Rules (d) and (h).*

(g) Scales. The scales should be placed at the left and at the bottom of the graph; sometimes the vertical scale is repeated on the right (Fig. 38).

(h) Units. Units should always be indicated at the top of the y-axis. The units relating to the horizontal scale are indicated as shown in Figs. 35, 37 and 38.

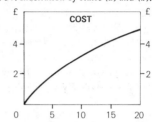

Fig. 38. *Illustration of Rules (g) and (h).*

(i) Curves. When more than one curve occurs on the same graph, these must be clearly distinguished.

The most usual methods are:

(i) differentiated lines, e.g. ── ── ── ── ── ── ·········· ──·── ·──·──.

(ii) different colours.

Both these methods necessitate a key.

(iii) using captions as shown.

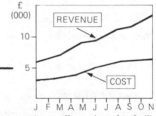

Fig. 39. *Illustration of Rule (i).*

(j) Too many curves should not be included in one graph.

(k) Title. The graph must be given a title. This must be clear and concise. Where necessary, sub-titles and footnotes are added.

(l) Frame. Often the graph is completely framed to give a finished appearance.

EXAMPLES OF GOOD GRAPHS

Figures 40 and 41 are excellent examples of the way in which graphs should be made. They are taken from *Economic Trends*, published by HMSO. Note the differentiated lines to distinguish the years, the repetition of the vertical scale at the right in order to facilitate the reading of the graphs, how attention is directed to the fact that the space between the zero lines and the first reading on the vertical scale is curtailed so that a false impression is not given. Note further sub-titles. In the case of Fig. 40, relating to production, note that the points are plotted in the middle of the spaces relating to each month, whereas in the case of the graph relating to stocks the points are plotted at the end of the space relating to each month.

Whether production or sales are plotted in the middle of the spaces allotted to the periods to which they refer or whether they are plotted as in Fig. 33 is a matter of choice.

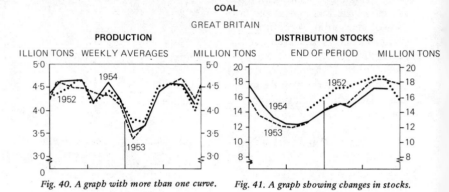

COAL

GREAT BRITAIN

PRODUCTION   DISTRIBUTION STOCKS

ILLION TONS   WEEKLY AVERAGES   MILLION TONS   END OF PERIOD   MILLION TONS

Fig. 40. A graph with more than one curve.   Fig. 41. A graph showing changes in stocks.

TYPES OF GRAPH

THE Z CHART

This chart consists of three curves on a single chart. When drawn, the figure resembles the letter Z, hence its name. It is used to show movements in sales, production, etc.

The three curves arc in respect of:

(a) the original data relating to each week, month or quarter as the case may be,

(b) the cumulative amount, and

(c) the moving annual total.

The scales used for the cumulative and annual total are usually smaller than the scale of the original data. This enables the movements of the original data to be shown more clearly. In the case of monthly figures, five times is a suitable amount. In the case of weekly figures, twenty times would be suitable. It is usual to differentiate the curve relating to the original data from the cumulative and the moving total curves by using different coloured lines. The cumulative curve always meets the moving total curve at the end of the year.

The curve relating to current data will show the fluctuations of a

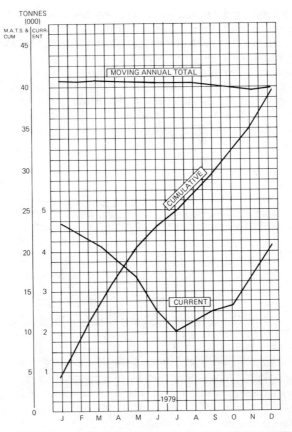

	Month	Monthly (tonnes)	Cumulative total to date (tonnes)	Moving annual total (tonnes)
1979	January	4,666	4,666	40,666
	February	4,416	9,082	40,582
	March	4,166	13,248	40,748
	April	3,966	17,214	40,714
	May	3,333	20,547	40,547
	June	2,500	23,047	40,547
	July	2,003	25,050	40,550
	August	2,247	27,297	40,297
	September	2,512	29,809	40,309
	October	2,688	32,497	39,997
	November	3,330	35,827	39,827
	December	4,169	39,996	39,996
1980	January	4,700	4,700	40,030
	February	4,515	9,215	40,129
	March	4,410	13,625	40,373

Fig. 42. A Z chart.

seasonal nature, the cumulative curve shows the position to date, and the moving total curve shows the trend.

A Z chart together with the data from which it is constructed is shown in Fig. 42.

The year is from January to December, The cumulative totals will therefore re-commence each January. The cumulative total is the total to date; the cumulative total for April 1979 will therefore be 4,666 + 4,416 + 4,166 + 3,966 = 17,214 as given in the table. The cumulative total for May 1979 will be 17,214 + 3,333 = 20,547,

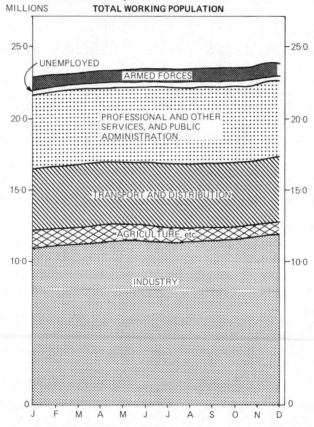

Fig. 43. A band curve chart showing composition of total working population by employment.

again as given in the table. It is not possible to check the moving Annual Total for 1979 as the previous year's monthly figures are not given. However, it is possible to check them for the three months of 1980. The moving total for December 1979 is 39,996, this being the total for the previous twelve months. To find the moving total for January 1980, the monthly figure for the previous January is deducted and the figure for January 1980 is added, thus 39,996 − 4,666 + 4,700 = 40,030. For February 1980 the total for the twelve months ended January 1980 is taken, the production of the previous February subtracted and the production for February 1980 added, thus 40,030 − 4,416 + 4,515 = 40,129.

BAND-CURVE CHARTS

In these charts, the constituent parts of the whole are plotted one above the other, so that the charts appear, when shaded, to consist of a series of bands.

The whole space between the top curve and the base is filled in solidly (Fig. 43). Band-curve charts can also be made where the data are put into percentage form; the whole chart will depict 100 per cent and the bands the percentage each component bears of the whole.

THE LORENZ CURVE

This curve is a graphic method of showing to what extent various magnitudes vary from uniformity.

Examples of the use of such a curve are:

(a) to show the degree of concentration of a particular industry in the hands of a few firms;

(b) the extent to which the incomes of various groups of people vary;

(c) the extent to which the profits of various firms vary.

The basis of the curve is the plotting of the cumulative percentage of the number of items, e.g. firms or people against the cumulative percentage of, for example, the sales, incomes or profits distributed among those items.

If all the firms in a particular industry were the same size (measuring size by their sales), then 10 per cent of the firms would have 10 per cent of the sales, 30 per cent of the firms would have 30 per cent of the sales, and so on. Plotting cumulative percentage sales against cumulative percentage number of firms would result in a straight line, and since the chart is square, the line would be at an angle of 45°. This is the line of equal distribution and is always drawn. The more the actual curve diverges from this line the greater the divergence from a uniform distribution.

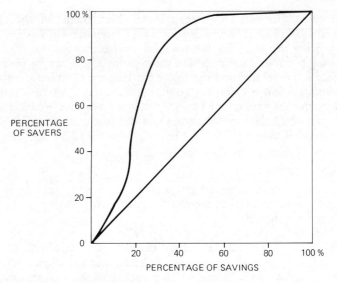

Annual savings	Fre-quency	Amount saved	Cumulative savings		Cumulative frequency	
		£	£	%		%
Not exceeding 50p	9,133	2,511	2,511	29	9,133	80
50p to £1.00	1,503	998	3,509	41	10,636	93
£1.00 to £5.00	546	1,002	4,511	53	11,182	98
£5.00 to £15.00	95	971	5,482	64	11,277	99
£15.00 to £35.00	51	1,024	6,506	76	11,328	99
£35.00 to £100.00	22	985	7,491	89	11,350	99
Over £100.00	5	1,013	8,504	100	11,355	100
	11,355	8,504				

Fig. 44. A Lorenz curve.

PROCEDURE

Drawing a Lorenz curve.

1. Draw up a frequency table as in the first two columns of the table in Fig. 44.
2. Compute the amount of sales, profits, incomes as the case may be for each group. In Fig. 44, this is savings and is given in column 3.

3. Compute the cumulative amounts, in the example shown, savings and the cumulative frequencies (columns 4 and 6 in the table in Fig. 44).
4. Compute the cumulative percentage amounts and the cumulative percentage frequencies (columns 5 and 7 in the table in Fig. 44).
5. Plot cumulative percentage frequencies against cumulative percentage amounts.
6. Draw the 45° diagonal.

Note. The scales are the same for both axes.

Interpretation of a Lorenz curve. Figure 44 shows that the greater amount of savings is done by comparatively few people and that a large number of people only save a comparatively small amount of the total. In other words, there is a great divergence between the amounts saved by the various savers.

If more than one curve is plotted on the same chart, it is possible to compare, say, whether teachers' salaries vary more than those of bank clerks, or whether there is a greater concentration of trade in fish shops or butchers' shops. The farther away from the diagonal, the greater the amount of concentration. It is also possible to make comparisons between areas, or at different times.

It is, of course, equally possible to make the comparisons by comparing different charts. However, it is easier if the curves are on the same chart.

QUESTIONS

1. From the sales of Puro-Products Ltd. given below, draw a Z chart.

	1978 (£000s)	1979 (£000s)		1978 (£000s)	1979 (£000s)
January	143	156	July	208	306
February	172	182	August	135	214
March	181	201	September	118	239
April	209	234	October	241	337
May	412	512	November	326	385
June	413	511	December	389	411

2. From the data given below, draw a band-curve chart.

Road Casualties

	Total	Pedes-trians	Pedal cyclists	Motor-cyclists and passengers	Other drivers and passengers
1973	216,493	59,861	48,077	42,680	65,875
1974	208,012	54,503	47,244	42,644	63,621
1975	226,770	58,553	49,618	49,438	69,161
1976	238,281	61,381	49,295	52,531	75,074
1977	267,922	63,897	52,447	63,602	87,976

3. State briefly, giving reasons, the kind of diagram you consider most appropriate for use with each of the following classes of statistical data:

(*a*) Number of children per family in a large town.

(*b*) Monthly rainfall for a period of three years.

(*c*) Total ship canal dues for one year according to country of registration of ships using the canal.

(*d*) Noon temperature for each day for six months at one weather station.

(*e*) Monthly output of steel for one year according to the principal grades of quality.   *Chartered Institute of Transport.*

4. What is meant by a historigram? From the data below, draw a historigram.

Gross National Product

£ million

1970	.	.	8,787	1975	.	.	12,793
1971	.	.	9,387	1976	.	.	13,928
1972	.	.	10,376	1977	.	.	14,858
1973	.	.	11,057	1978	.	.	15,909
1974	.	.	11,636	1979	.	.	16,784

5. The Managing Director of a firm engaged in the manufacture of men's shirts asks you to produce information in the form of tables or graphs, which he wishes to see as a matter of routine. Draft four specimen tables or graphs stating how often they would be compiled and what information he would derive from each of them.

Institute of Statisticians.

6. What does a graph show? On which axis is the independent variable plotted?

7. Draw up a list of rules for the construction of graphs.

8. The following figures come from a Report on the Census of Production.

Textiles machinery and accessories

Establishments (Nos.)	Net output (£000)
48	1,406
42	2,263
38	3,699
21	2,836
26	3,152
16	5,032
23	20,385
214	38,773

Analyse this table by means of a Lorenz curve and explain what this curve shows. *Institute of Cost and Management Accountants.*

CHAPTER 12

Ratio-Scale Graphs

The graphs in the preceding chapter were concerned with movements in absolute quantities, e.g. an increase or a decrease of so many tonnes or so many £s. In the present chapter the movements that are to be shown are not the absolute movements, but the rates of change of the movements, i.e. relative changes.

RELATIVE AND ABSOLUTE CHANGES

The difference is best explained by means of an example.

Consider a retail shop whose sales are in four successive weeks as follows:

$$£125 \qquad £150 \qquad £180 \qquad £216.$$

Successive increases are:

$$£25 \qquad £30 \qquad £36$$

These are absolute quantities.

Percentage increases are as follows:

$$\frac{25 \times 100}{125} \qquad \frac{30 \times 100}{150} \qquad \frac{36 \times 100}{180}$$

These are relative changes. It will be noted that the increase in sales over the preceding week is in each case 20 per cent. If the original data were plotted on semi-logarithmic paper, the result would be a straight line, showing that the rate of increase was constant. On ordinary natural-scale graph paper absolute differences would be shown and it would not be possible to say anything about the rate of change.

THE USES OF SEMI-LOGARITHMIC SCALE GRAPHS

If the interest is in the rate of change of data, and not in the actual absolute changes, the data must be plotted on semi-logarithmic paper; such graphs are known as ratio-scale graphs or semi-logarithmic graphs.

Here are some of the uses of such a graph.

(a) Comparison of a firm's sales with growth of total sales of the industry to see if the firm is obtaining its share of sales.

(b) Comparison of the efficiency of salesmen.

(c) Comparison of outputs of various departments.

(d) Estimation of future sales.

(e) Comparison of the fluctuations of prices of different commodities.

HOW TO USE SEMI-LOGARITHMIC PAPER

On examining semi-logarithmic paper, a number of differences from ordinary natural-scale paper will be noticed: there is no zero line on the vertical scale, the lines from 1 to 10 become closer and closer, the pattern is repeated, but the values are ten times greater. If the pattern were repeated a third time, i.e. if the paper had three cycles, the third cycle would be from 100 to 1,000, i.e. ten times greater than the second cycle. Similarly, the fourth cycle would have values ten times greater than the third, and so on. Alternatively, the first cycle could have started with the value of 0.1 and finished at 1, in which case the second cycle would have values from 1 to 10, the third from 10 to 100 and so on. Note that the units are sub-divided into ten parts when the lines are fairly far apart and into 5, when they are closer together. Note further that the sub-divisions also become closer and closer. This scale is logarithmic. The horizontal scale is the ordinary arithmetical scale. Hence the paper is known as semi-logarithmic.

Actual values are plotted on this paper. If semi-logarithmic paper is not available, however, the logarithms of the values can be plotted on arithmetically scaled paper, but the actual values are inserted on the scale. This is, in effect, making a logarithmic scale.

The reason for the scale on semi-logarithmic paper starting at 1, and not zero, is that the logarithm of 1 is 0; hence the value of 1 is placed at zero distance from the origin, i.e. at the origin. There is no logarithm for zero, nor for negative numbers, hence such values cannot be plotted.

Students should carefully note exactly how the values given are plotted on the ratio-scale graph in Fig. 45.

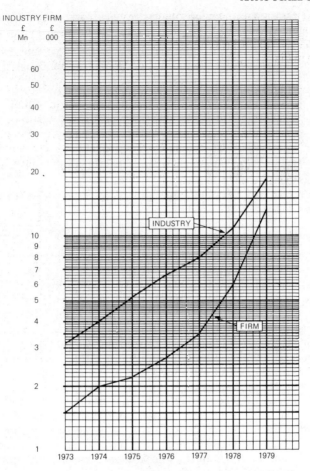

	Sales	
	Firm (£000)	*Industry* (£m)
1973	1.50	3.2
1974	2.00	4.0
1975	2.18	5.4
1976	2.69	6.8
1977	3.50	8.0
1978	6.00	11.0
1979	14.00	19.0

Fig. 45. Comparison of relative movements on ratio-scale graphs.

In order to make comparisons easier when comparing the rates of movement of two quantities, the graphs can be moved up or down the chart so as to bring them closer together. This can be done as shown in Fig. 45. Great care is needed in this "moving" of the curves on a logarithmic chart. It can be done by changing the value of the units on the scale. Thus, in the example given, the first cycle shows values 1 to 10 in thousands of £s for the firm, but in millions of £s for the industry. If this were not done, one curve would be three cycles away from the other, making comparison more difficult.

Finally, it should be noted that the position of the curves has no meaning. Their meaning is derived from their direction, changes in direction and their degree of steepness.

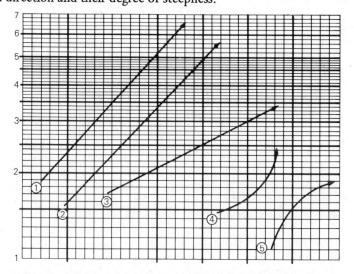

Fig. 46. Movements on semi-logarithmic graphs.

MEANING OF LINES ON SEMI-LOGARITHMIC GRAPHS

Figure 46 shows a number of lines drawn on ratio-scale graph paper. Their meaning is as follows:

(*a*) 1 and 2 are increasing at the same rate, since they are parallel. They are both increasing at a constant rate, since they are both straight.

(*b*) 3 is also increasing at a constant rate, but not so quickly as 1 or 2, since this curve is less steep.

(*c*) 4 is increasing at an increasing rate, since it is becoming steeper and steeper.

(*d*) 5 is increasing at a decreasing rate, since the curve is becoming less and less steep.

A COMPARISON OF RATIO-SCALE AND ARITHMETICAL-SCALE GRAPHS

A comparison can be made of these two types of graphs as follows:

Ratio-Scale

(*a*) Used to show relative movements.

(*b*) Equal vertical distances represent equal proportional changes.

(If 1 centimetre represents an increase of 40 per cent in one part of the graph, then 1 centimetre represents 40 per cent in any other part. However, ½ centimetre does not represent 20 per cent.)

(*c*) Zero and negative quantities cannot be shown.

(*d*) No base line necessary. Graph can be moved up and down without changing meaning.

(*e*) Meaning derived from direction of lines.

(*f*) Cannot be used to show component parts.

(*g*) Two or more series of entirely different units, e.g. tonnes, litres, £s can be compared on the same graph.

(*h*) A very great range of values can be shown.

Arithmetical-Scale

(*a*) Used to show movements of absolute quantities.

(*b*) Equal vertical distances represent equal amounts.

(If 1 centimetre represents £4, then 1 centimetre represents £4 in any other part of the graph. ½ centimetre will represent £2.)

(*c*) Zero and negative quantities can be shown.

(*d*) Base line necessary. Graph cannot be moved up or down without destroying its meaning.

(*e*) Meaning derived from position of lines.

(*f*) Can be used to show the components (band-curve charts)

(*g*) It is not possible to show two or more series on the same graph for purposes of comparison if they are of entirely different units.

(*h*) Cannot show a very great range of values if the graph is to be clear.

QUESTIONS

1. Discuss the advantages and disadvantages of logarithmic charts. Describe and illustrate two practical applications of this type of diagram.

2. The following table shows the number of television licences current at the end of the stated periods.

							000's
1948		.	.	.	.	.	93
1949		.	.	.	.	.	239
1950		.	.	.	.	.	578
1951	March.		.	.	.	.	764
	June .		.	.	.	.	897
	September	.		.	.	.	949
	December	.		.	.	.	1,162
1952	March.		.	.	.	.	1,456
	June .		.	.	.	.	1,539
	September	.		.	.	.	1,655
	December	.		.	.	.	1,839
1953	March.		.	.	.	.	2,142
	June .		.	.	.	.	2,415
	September	.		.	.	.	2,615

Plot the series on a suitable graph and estimate the figure as at end December 1953. Would this series be useful for the purpose of interpreting trends in passenger traffic?

Chartered Instituted of Transport.

3. Discuss the limitations and uses of graphs in statistical analysis. State the respective merits of *(a)* natural-scale graphs; *(b)* semi-logarithmic graphs; *(c)* bar charts.

4. Explain what is meant by a semi-logarithmic diagram and discuss its advantages over the natural-scale diagram.

Graph the two following series on the same diagram so as to bring out their relative movements.

New Factory Buildings in Patonia
Number Completed

				Total	No. in development areas
1969	1	.	.	264	92
	2	.	.	267	122
	3	.	.	215	48
	4	.	.	276	90
1970	1	.	.	256	58
	2	.	.	284	65
	3	.	.	218	48
	4	.	.	264	48

1971	1	.	.	241	59
	2	.	.	248	65
	3	.	.	271	46
	4	.	.	279	51

London Chamber of Commerce and Industry.

5. What are the respective merits of *(a)* arithmetic charts, *(b)* semi-logarithmic charts and *(c)* bar charts, in presenting statistics showing *(i)* sales and *(ii)* stocks of a number of different products? Indicate, with reasons, which method you regard as best for each set of statistics.   *Institute of Statisticians.*

6. State the chief points to be considered in the construction of *(a)* the arithmetic-scale graph and *(b)* the ratio-scale graph. Graph the data below using a ratio-scale graph.

Road Accident Deaths

	1975	1976	1977	1978	1979
Number of deaths per million of population . .	106	102	93	98	102
Consumption of motor fuel in 10 million gallons .	35	31	30	29	27

7. Below are given estimates of world population at various dates. From this information make an estimate of the population for the year 2020, using graphical means.

Year	1670	1770	1870	1920	1970
Population (millions)	470	691	1,091	1,550	2,406

8. Using only the following information:

$$\log 2 = 0.301, \log 3 = 0.477$$
$$\log 7 = 0.845$$

illustrate clearly how you would rule some semi-logarithmic paper, using ordinary graph paper. Show fully how you would insert scales on both axes.

Institute of Cost and Management Accountants.

CHAPTER 13

Frequency Distributions

VARIABLES

The price of a commodity can vary from shop to shop; height will vary from person to person; the number of children will vary from family to family. Prices, heights and the number of children per family are examples of a *variable* (also termed a *variate*). The various prices, heights and numbers of children are the values of the variable.

Variables are of two kinds:

(a) continuous variables — those which can take any value, e.g. heights.

(b) discrete variables — those which can only take particular values, e.g. the number of children in a family (it is not possible to have two and a half children).

The values of continuous variables are obtained by measurement; the values of discrete by counting.

FREQUENCY TABLES

The following is an example of a *frequency table*. It shows the *frequency distribution* of the heights of a group of students.

Example 1.

Height (metres)	Number of students
1.50 - 1.60	18
1.60 - 1.70	36
1.70 - 1.80	27
1.80 - 1.90	9
	90

This table shows that 18 students have heights that vary between 1.50 metres and 1.60 metres; that 36 students have heights that vary between 1.60 metres and 1.70 metres, and so on.

The height is the *variable*; the number of students is the *frequency*. In the example there are four *classes*, namely 1.50 - 1.60; 1.60 - 1.70; 1.70 - 1.80; and 1.80 - 1.90. The *lower limit* of the first class is 1.50 metres, the *upper limit* is 1.60 metres and the *class interval* is 0.10 metres (the difference between the upper and lower limits). The lower limit of the second class is 1.60 metres, the upper limit 1.70 metres and again the class interval is 0.10 metres. It will be observed that the upper limit of one class is the same as the lower limit of the following class. These limits are known as *mathematical limits*.

Note. In all cases mathematical limits must be used for calculations and for making graphs of the distribution.

MATHEMATICAL LIMITS OF CLASS INTERVALS

Usually frequency tables are not shown with mathematical limits. The table below shows how this matter is dealt with.

How presented	Assumptions	Mathematical limits
(1. area: hectares)		
5 -		
10 -		5 - 10
15 -		
or		10 - 15
5 but under 10		
10 ” ” 15		15 - 20
15 ” ” 20		
5 - 9	that the areas of the	4.5 - 9.5
10 - 14	fields have been	9.5 - 14.5
15 - 19	recorded to the	14.5 - 19.5
	nearest hectare.	
(2. ages: years)		
0 - 4	the age is recorded	0 - 5
5 - 9	as age last birthday	5 - 10
10 - 14	however near to the	10 - 15
	next.	

COMPILING FREQUENCY TABLES

The tables are compiled from specific measurements, amounts or quantities. From this mass of data the statistician will have to decide (a) the number of classes, and (b) the class intervals.

(a) *The number of classes.* If too many classes are decided upon certain of them may have too few frequencies and the distribution may be shown as more irregular than it really is. On the other hand, if too few classes are chosen, the distribution will be shown less accurately than is necessary. This point will be appreciated when frequency distributions are shown graphically, later in this chapter. The more items there are to be grouped, the greater the number of classes. A maximum of 20 classes when there is a very large amount of data, and a minimum of, say, 5 when the number of items is small might be considered a practical rule.

(b) *The class intervals.* What really has to be decided here are the class limits. Unnecessary difficulties will be avoided if the class limits do not coincide with any recorded values and since it is necessary to treat all the values in any one class as if they were all equal to the mid-value of the interval, class limits are chosen so that values around which a large number of items occur tend to be the mid-values of intervals. If at all possible, class intervals should be made equal. In Example 1, the 18 students in the first class are assumed to have an average height of 1.55 metres; in the second class 1.65 metres, and so on; and the class intervals are equal.

HISTOGRAMS

Frequency tables can be presented in the form of a diagram known as a histogram. This consists of a series of rectangles having a *base* measured along the *x*-axis *proportional to the class interval* and an *area proportional to the frequency.* The heights of the rectangles will be the frequency (the area) divided by the class interval (the width). The height of the rectangles measure the *frequency density.*

Where the class intervals are equal, the heights of the rectangles are proportional to the frequencies. Nevertheless, the heights still measure the frequency density and the areas are proportional to the frequencies.

The frequency table of Example 1 is shown in the form of a histogram in Fig. 47. In this case the class intervals are equal and so the heights of the rectangles are proportional to the frequencies which are denoted on top of the rectangles.

Note. Do not show the frequencies along the *y*-axis. Frequency density is measured along the *y*-axis, as is shown in Fig. 48 which uses the data from Example 2.

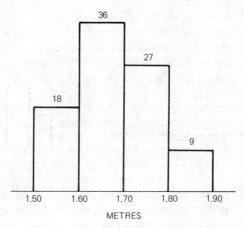

Fig. 47. *Histogram where class intervals are equal.*

Example 2.

Capacity (centilitres)	Number of bottles	Class interval (centilitres)	Frequency density (bottles per cl)
0 - 8	16	8	$\frac{16}{8} = 2$
8 - 17	54	9	$\frac{54}{9} = 6$
17 - 24	21	7	$\frac{21}{7} = 3$
24 - 33	9	9	$\frac{9}{9} = 1$

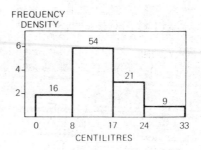

Fig. 48. *Histogram where class intervals are unequal.*

Example 3.

Number of lorries	Number of operators	Frequency density
1	7	7
2	13	13
3	9	9
4 - 5	6	3
	35	

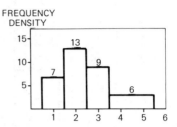

Fig. 49. Histogram where the variable is discrete.

Bearing in mind that mathematical limits are necessary for each class and that the mid-value of the interval should be equal to the value of the variable, the class limits are:

½ - 1½;   1½ - 2½;   2½ - 3½;   3½ - 5½.

The histogram for this distribution is shown in Fig. 49.

PROCEDURE

Drawing a histogram

1. Ensure that you have mathematical limits.
2. Compute class intervals.
3. Compute frequency densities (frequency ÷ class interval).
4. Decide scales for *(i)* the variable, and *(ii)* the frequency density.
5. Draw histogram: height of rectangles = frequency density; width of rectangles = class interval.

Note. A histogram is *not* a bar chart. The area of a bar chart has no meaning; the area of a histogram has.

FREQUENCY POLYGONS

When the class intervals of a frequency distribution are equal, it can also be represented by means of a frequency polygon. This is obtained by joining the mid-points of the tops of the rectangles of a histogram. An example is shown in Fig. 50. (The corresponding histogram is shown in Fig. 47.) Notice the treatment of the first and last rectangles. The areas of the frequency polygon and the histogram from which it was derived are thus the same.

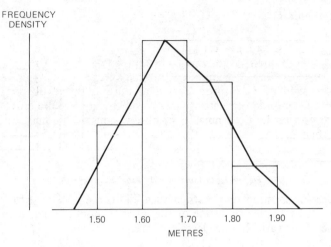

Fig. 50. A frequency polygon.

FREQUENCY CURVES

As the class intervals become smaller and smaller and the number of observations becomes greater and greater, the frequency polygon and the histogram approach a smooth curve. This process, carried to the limit, would result in a frequency curve.

THE SMOOTHED HISTOGRAM AS A FREQUENCY CURVE

The histogram assumes, usually quite unrealistically, that over the class intervals different magnitudes have the same frequencies. For this reason, a frequency curve is more accurate. An attempt to obtain such a curve can be made by smoothing the histogram. An example is shown in Fig. 51.

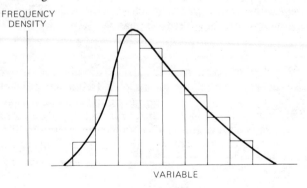

Fig. 51. A "smoothed" histogram.

THE MEANING OF A FREQUENCY CURVE

Figure 52 shows a frequency curve. The total area under the curve is proportional to the total number of observations, i.e. the total frequency. The shaded area is proportional to the number of observations whose magnitude varies between Z_1 and Z_2. The area to the right of Z_2 is proportional to the number of observations whose magnitude is greater than Z_2 and the area to the left of Z_1 is proportional to the number of observations whose magnitude is less than Z_1.

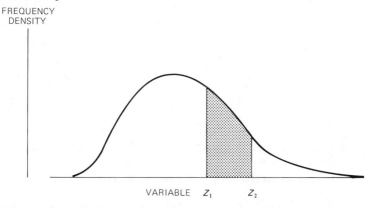

Fig. 52. A frequency curve.

FUNDAMENTAL TYPES OF FREQUENCY DISTRIBUTIONS

(a) The symmetrical distribution. The class frequencies decrease to zero symmetrically on either side of the central maximum. Figure 53 shows the bell-shaped frequency curve for this type of distribution. An example of such a distribution would be the heights of adult males or adult females.

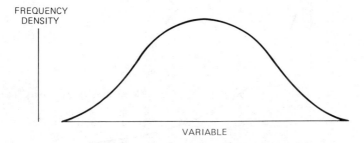

Fig. 53. Symmetrical frequency curve.

(b) *The moderately asymmetrical or skewed distribution.* The class frequencies decrease with greater rapidity on one side of the maximum than on the other. Figure 54 shows the frequency curves of two such distributions, (a) being positively skewed, (b) being negatively skewed. This type of distribution is very frequent. An example of a positively skewed distribution is the number of marriages classified according to the age of the bridegroom. An example of a negatively skewed distribution, given by Yule and Kendall, is the distribution of barometric heights at Greenwich.

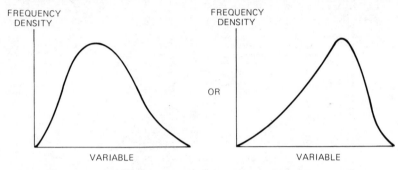

Fig. 54. *Asymmetrical frequency curve (a) positively skewed; (b) negatively skewed.*

(c) *The J-shaped distribution.* This is an extremely asymmetrical distribution: the class frequencies are highest at one end of the range. Figure 55 shows a frequency curve of a J distribution where the highest frequencies are at the lower end of the range. An example of such a distribution is the distribution of incomes.

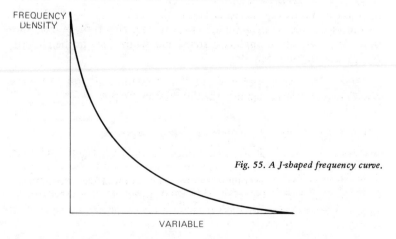

Fig. 55. *A J-shaped frequency curve.*

CUMULATIVE FREQUENCY CURVES

A frequency distribution can also be presented in the form of a cumulative frequency curve (sometimes called an Ogive, sometimes, erroneously, a cumulative frequency polygon).

Example 4.

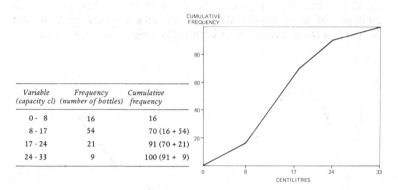

Variable (capacity cl)	Frequency (number of bottles)	Cumulative frequency
0 - 8	16	16
8 - 17	54	70 (16 + 54)
17 - 24	21	91 (70 + 21)
24 - 33	9	100 (91 + 9)

Fig. 56. A cumulative frequency curve.

This means that 16 bottles have capacities which do not exceed 8 cl, 70 bottles which do not exceed 17 cl, 91 bottles which do not exceed 24 cl and 100 which do not exceed 33 cl.

The cumulative frequency curve is drawn by plotting the cumulative frequencies against the upper mathematical limits of the class intervals as shown in Fig. 56.

It will be noted that the unequal class intervals cause no difficulty. The points that have been plotted have been joined by straight lines. This is usual, although inaccurate; a smooth curve should pass through the points.

Note. It is the upper limit of the class interval (*not* the midpoint) that is plotted against the cumulative frequency.

PERCENTAGE CUMULATIVE FREQUENCY CURVES

For these curves, it is the percentage cumulative frequencies that are plotted against the upper limits of the class intervals. The percentage cumulative frequencies are the cumulative frequencies expressed as percentages of the total number of observations. Figure 57 shows the percentage cumulative frequency curve for the data given in Example 5.

Example 5.

Variable (height: metres)	Frequency (number of students)	Cumulative frequency	Percentage cumulative frequency
1.50 - 1.60	18	18	$\frac{18}{90} \times 100 = 20$
1.60 - 1.70	36	54	$\frac{54}{90} \times 100 = 60$
1.70 - 1.80	27	81	$\frac{81}{90} \times 100 = 90$
1.80 - 1.90	$\frac{9}{90}$	90	$\frac{90}{90} \times 100 = 100$

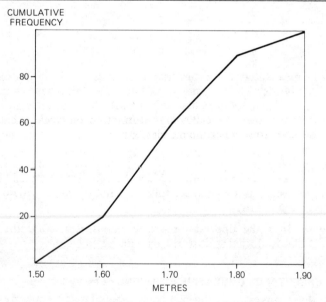

Fig. 57. A percentage cumulative frequency curve.

Note. When it is necessary to compare two or more cumulative frequency curves on the same diagram, they *must* be percentage cumulative curves.

PROCEDURE

Drawing a cumulative frequency curve

1. Ensure that you have mathematical limits.
2. Compute cumulative frequencies.
3. Plot cumulative frequencies against upper limits of class intervals and plot zero frequency against lower limit of the first class.
4. Join points plotted by straight lines.

Drawing a percentage cumulative frequency curve

1. Ensure that you have mathematical limits.
2. Compute cumulative frequencies.
3. Compute percentage cumulative frequencies (i.e. cumulative frequencies as a percentage of the total number of observations).
4. Plot percentage cumulative frequencies against upper limits of class intervals and plot zero cumulative frequency against lower limit of the first class.
5. Join points plotted by straight lines.

QUESTIONS

1. A person's socio-economic status can be classified as either A, B, C1, C2, D or E in descending order. A random sample of 60 individuals gave the following information on weekly earnings (£) in relation to socio-economic class:

45(B)	20(E)	16(D)	61(C2)	18(E)
32(C2)	22(C2)	64(C1)	62(C1)	33(D)
49(D)	28(C1)	60(C2)	74(B)	50(C2)
49(C2)	64(B)	33(D)	29(C2)	21(E)
24(C1)	48(C2)	23(D)	27(C2)	26(E)
26(D)	37(C2)	85(A)	18(C1)	42(C2)
43(C1)	37(C2)	67(B)	19(E)	22(D)
17(C2)	17(C2)	19(D)	23(D)	50(C2)
66(C1)	74(C1)	17(E)	37(C2)	55(C1)
52(C2)	37(C2)	26(D)	42(C1)	40(C2)
79(B)	23(E)	24(D)	31(D)	44(C1)
15(E)	18(E)	40(C2)	65(C1)	17(E)

Required:

(a) Compile a frequency distribution of earnings with intervals of a suitable width.

(b) By considering earnings to fall into one of three groups of less than £25, £25 - 45, and more than £45, compile a two-way table showing the frequencies in each earnings group/socio-economic class combination.

(c) Describe the main features of the data as observed from the table compiled in answer to *(b)* above.

The Association of Certified Accountants.

2. Comment on the suitability of the class intervals below for use in connection with data, of which a sample item would be 29.9.

A	B	C
0 - 10	0 - 9.9	0 - 9.95
10 - 20	10 - 19.9	9.95 - 19.95
20 - 30	20 - 29.9	19.95 - 29.95
30 - 40	30 - 39.9	29.95 - 39.95

3. The frequency distribution below shows the number of marks obtained in a recent examination.

Marks given	Number of examinees
0 - 10	2
11 - 20	3
21 - 30	11
31 - 40	23
41 - 50	41
51 - 60	67
61 - 70	29
71 - 80	17
81 - 90	5
91 - 100	2
	200

Draw a histogram of the distribution and on the same chart draw a frequency polygon.

4. Below are the wage distributions of two firms, Exo and Zedo. Compare them by drawing percentage cumulative frequency curves.

Weekly wages £	Number of employees Exo	Zedo
20 and under 30	6	9
30 ” ” 40	19	33
40 ” ” 60	21	68
60 ” ” 80	26	71
80 ” ” 100	6	11
100 ” ” 120	2	8
	80	200

5. There are twenty general transport firms in the town of Cowston-on-Ryle. The lorry distribution is as follows:

Number of lorries	Number of firms
1	2
2	5
3	6
4	3
5 & 6	4

Draw a histogram for this distribution.

CHAPTER 14

Averages

An average is a value which is typical or representative of a set of values. It lies within a set of data arranged according to magnitude and hence it is *a measure of a central tendency*. It does not necessarily coincide with any particular value.

The averages to be dealt with in this chapter are:

(a) the arithmetic average (or arithmetic mean),
(b) the harmonic mean,
(c) the median,
(d) the mode,
(e) the geometric mean.

Note. When the word "mean" or "average" is used without any indication as to the kind of average or mean, the arithmetic mean is usually (but not necessarily) the one intended.

(A table comparing the averages is given at the end of the chapter in Fig. 63.)

THE ARITHMETIC MEAN

This is found by dividing the sum of the values of the variable by the number of values.

Example 1

Consider five pieces of string having the following lengths: 4 cm, 8 cm, 10 cm, 11 cm, and 12 cm (i.e. 5 values). The arithmetic mean is calculated thus:

$$(4 + 8 + 10 + 11 + 12)/5 = 9 \text{ cm.}$$

Usually there is more than one item of a given value; that is, for each value there is a frequency. To find the arithmetic mean each value is multiplied by the relevant frequency. These products are then added and the sum divided by the total number of values, that is the total of the frequencies.

Example 2

There are nine pieces of string. The lengths are as follows:

1 cm . . . 1 piece,   2 cm . . . 2 pieces,   3 cm . . . 1 piece,
5 cm . . . 2 pieces,   6 cm . . . 3 pieces.

The mean length will be calculated thus:

$$\frac{(1)(1) + (2)(2) + (3)(1) + (5)(2) + (6)(3)}{1 + 2 + 1 + 2 + 3} \text{ cm} = 4 \text{ cm}.$$

This method of finding the mean has just been described and the carrying out of the method has been shown in Examples 1 and 2. However, by using symbols, the description of the method to find the mean can be shortened, thus:

$$\bar{x} = \frac{\Sigma fx}{N}.$$

$\bar{x}$ (pronounced "x bar") stands for the arithmetic mean.

m is also used to denote the arithmetic mean.

x is the value of the variable.

f is the frequency.

fx is the product of the value of the variable and the relative frequency.

Σ (the Greek capital letter sigma) means "add up all the quantities like . . .". In this case all the fx's.

N is the total number of observations (i.e. values) which equals the sum of the frequencies: $N = \Sigma f.$

Note. Mathematical symbols do not necessarily imply advanced mathematics. They are an exceedingly convenient form of shorthand. If ordinary language were used instead, a considerable amount of time would be wasted. Any difficulty students may encounter at first is due to their unfamiliarity and will soon disappear.

WEIGHTED ARITHMETIC AVERAGE

The mean length of the nine pieces of string in Example 2 was a weighted average. There were five different lengths and if these five lengths had been added up and divided by five the average length obtained would have been $(1 + 2 + 3 + 5 + 6)/5 = 3.4$ cm. This is an unweighted average, more correctly described as an equally weighted average. It is, however, incorrect; no *importance* or *weight* is given to the larger number of pieces of 6 cm. When this is taken into consideration together with all the other frequencies, as it *must* be, the effect is to increase the figure of 3.4 cm to 4 cm, the correct arithmetic mean.

An analogy taken from the world of physics will help show the nature of an arithmetic mean. Figure 58 shows a balance. The arm which is graduated is balanced around the fulcrum which is marked F and which is at graduation 4. From the arm are suspended weights. The system is in equilibrium. If more weights were added to those on the left of the fulcrum or existing weights were suspended at a greater distance from the fulcrum, the system would no longer balance and the arm would fall on the left of the fulcrum and incline on the right.

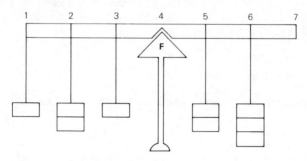

Fig. 58. The mean as the balancing value of a frequency distribution.

The system is in equilibrium because the forces trying to pull the arm down on the left side are equal to those trying to pull it down on the right. The force exerted downwards in respect of each weight is the weight multiplied by the distance from the fulcrum.

Forces pulling downwards on the left of the fulcrum:

$$(3)(1) + (2)(2) + (1)(1) = 8$$

Forces pulling downwards on the right of the fulcrum:

$$(2)(1) + (3)(2) = 8$$

It will be noticed that the weights are proportional to the frequencies in Example 2; that they are suspended from the arm at distances corresponding to the values of the variable; and that the fulcrum is at the graduation corresponding to the mean. The mean may there-

fore be considered as the value around which all the other values of
a frequency distribution are balanced.

THE PROPERTIES OF THE ARITHMETIC MEAN

Example 3.

x (length cm)	f (number of pieces)	d (x - $\bar{x}$)	fd	fd^2
1	1	−3	−3	9
2	2	−2	−4	8
3	1	−1	−1	1
5	2	1	2	2
6	3	2	6	12
	9		0	32

The first two columns of Example 3 repeat the data of Example 2
in columnar form. Column 3, headed by the symbol d, gives in
respect of each length the *deviation from the mean*. The mean is
known from Example 2 to be 4 cm. The deviations are obtained by
deducting from each value of the variable (i.e. from each x) the
mean value ($\bar{x}$) as follows:

x		$\bar{x}$		x - $\bar{x}$
1	−	4	=	−3
2	−	4	=	−2
⋮		⋮		⋮
6	−	4	=	2

As certain lengths occur more than once the deviations must be
multiplied by the relevant frequencies to obtain the total deviations.
This is done in the column headed fd. Column 5 (fd^2) gives the
result of squaring the deviation and multiplying by the relevant
frequency. This is effected by multiplying together columns 3 and 4
($d \times fd$).

Note. A frequent error is to square the figures in the fd column.
This would square the frequency as well, giving f^2d^2.

The total of the fd column is 0 (in symbols $\Sigma fd = 0$). The total of
the fd^2 column is 32 (in symbols $\Sigma fd^2 = 32$). Had the deviations
been taken from any other value than the mean, the total would
have been greater.

The properties of the arithmetic mean are therefore:
(a) $\Sigma fd = 0$;
(b) Σfd^2 is at a minimum.

THE ARITHMETIC MEAN COMPUTED FROM GROUPED DATA

When data are presented in the form of a frequency distribution *all values falling in a given class interval are considered as having the same value as the mid-point of the interval*, that is that the values have a mean equal to the mid-point.

Example 4.

x (£)	f	fx (£)	x + 1	f(x + 1)
20 - 40	12	360	31	372
40 - 50	28	1,260	46	1,288
50 - 60	60	3,300	56	3,360
60 - 80	15	1,050	71	1,065
80 and over	5	450	91	455
	N = 120	Σfx = 6,420		Σf(x + 1) = 6,540

The first two columns of Example 4 show a frequency distribution of grouped data. The last class has no upper limit. However, before the arithmetic mean can be calculated some reasonable assumption must be made about this *open-ended class.* The best guide to its probable limit is the pattern of the distribution. In the example £100 has been chosen as the upper limit. There are only a few items in this class compared with the total number of items and it would seem unlikely that any would exceed £100. The error will not therefore be excessive. The mid-point of the last class is therefore deemed to be £90.

The arithmetic mean is calculated £6,420/120 = £53.50.

CHARLIER'S CHECK

The last two columns of Example 4 provide a check on the arithmetic. In the column headed $x + 1$, the value of the variable is increased by 1 and the column headed $f(x + 1)$ gives the products of the frequencies and the increased values of the variable.

So, if the arithmetic is correct:

$$N + \Sigma fx = \Sigma f(x + 1)$$

which will be in the case of Example 4:

$$120 + 6,420 = 6,540.$$

PROCEDURE

Computing the arithmetic mean of grouped data

1. Ensure mathematical limits.
2. Make reasonable assumptions for any open-ended classes.
3. Calculate mid-points of the classes.
4. *(a)* Complete column *fx*.
   (b) If using Charlier's check, complete columns *x + 1* and *f(x + 1)*.
5. *(a)* Sum columns *f* and *fx*.
   (b) If using Charlier's check, sum column *f(x + 1)*.
6. *(a)* Substitute for *N* and Σ*fx* in the formula $\bar{x} = \Sigma fx/N$.
   (b) If using Charlier's check, verify $N + \Sigma fx = \Sigma f(x + 1)$.

THE HARMONIC MEAN

Care must be exercised in computing the mean when dealing with rates of the type *x per d* (such as kilometres per hour, pence per litre). Often the appropriate mean to use is the *harmonic mean*.

The harmonic mean is defined as *the reciprocal of the arithmetic mean of the reciprocals of the values.*

Example 5. Compute the harmonic mean of 2, 4, and 5.

$$\text{Harmonic Mean:} \frac{1}{\frac{1}{3}\left(\frac{1}{2}+\frac{1}{4}+\frac{1}{5}\right)} = \left(\frac{3}{\frac{19}{20}}\right) = 3.16.$$

The definition of the harmonic mean just given may also be expressed in symbols thus:

$$\text{Harmonic Mean:} \frac{1}{\frac{1}{N}\left(\frac{1}{x_1}+\frac{1}{x_2}+\ldots\frac{1}{x_N}\right)} = \frac{N}{\Sigma\left(\frac{1}{x}\right)}$$

Example 6. A cyclist rode from A to B, a distance of 120 kilometres, at a speed of 24 kilometres per hour and returned at a speed of 20 kilometres per hour.

Here the appropriate average is the harmonic mean:

$$\frac{2}{\dfrac{1}{24} + \dfrac{1}{20}} = 21\frac{9}{11} \text{ kilometres per hour.}$$

Check. The cyclist took 5 hours on the outward journey and 6 hours on the return. He took 11 hours to cover a total distance of 240 kilometres, that is $21\frac{9}{11}$ km per hour.

Suppose, however, the cyclist rode for an hour at 24 km per hour and then for another hour at 20 km per hour. His average speed would have been $(24 + 20)/2 = 22$ km per hour, the arithmetic mean.

It is thus seen that when averaging rates of the form *x per d* (in the example just given km per hour) sometimes the harmonic mean is appropriate and sometimes the arithmetic mean. In the first instance, the cyclist went the same distances but taking different times for the two journeys. The *x* of the *x per d* was constant. In the second case, the *d* (time taken) was constant.

Rules for averaging rates of the form *x per d* are as follows.

(a) The harmonic mean is used when rates are expressed in the form *x per d* and *x* is constant, or when they are expressed in the form *d per x* and *d* is constant.

(b) The arithmetic mean is used when rates are expressed in the form *x per d* and *d* is constant, or when they are expressed in the form *d per x* and *x* is constant.

It is worth remembering that the same rate can always be express-ed in two ways. *x per d* and *d per x*, e.g. 20 km per hour or 3 minutes per km.

THE MEDIAN

The arithmetic mean is not a satisfactory average for all purposes. If, for example, the average wage of a certain group of people was required, the type of average required would probably be one such that half of the group had a wage less than the average and the other half more than the average; a kind of half-way wage. This type of average is the *median*. It is a *positional* average.

The arithmetical mean would in the case of wages have been higher than the median, more people earning less than the arithmetical mean than those earning above. However, the median cannot be used for further calculations.

THE MEDIAN OF A DISCRETE SERIES

If all the values of the variable are put in order of magnitude (this is known as an *array*), the value of the middle item (the $(N + 1)/2$ item where N is the number of items) is the median.

Example 7.

Let 17g, 22g, 31g, 40g and 55g be an array.

There are 5 items; the *rank* of the median is therefore $(5 + 1)/2 = 3$. The value of the third item is 31g. This is the value of the median.

Example 8.

Let 2cc, 4cc, 5cc, 7cc, 8cc and 9cc be an array.

There are 6 items; the rank of the median is $(6 + 1)/2 = 3\frac{1}{2}$. The values of the 3rd and 4th items are averaged. The median is $(5 + 7)/2$ cc, that is 6 cc.

Note that in both Examples 7 and 8 there are an equal number of values above and below the median.

THE MEDIAN OF GROUPED DATA

In the case of a frequency distribution where the data is grouped the rank of the median is $N/2$. Its value is found by interpolation.

Example 9.

x	f	cum.f
£20 - 40	12	12
£40 - 50	28	40
£50 - 60	60	100
£60 - 80	15	115
£80 -	5	120

Rank of median: $\dfrac{120}{2} = 60$

Cum. frequency to £50: $\dfrac{40}{20}$

The "median" class:
interval £10
frequency 60

Median = $£(50 + \frac{20}{60} \times 10) = £53.3$

When finding the median of grouped data by interpolation, the *assumption* is made that the various *values* from the lower to the upper limit *in each class* are *distributed evenly* in order of size *thoughout the interval*.

In Example 9 it is assumed that the sixty values which are in the class containing the median increase by equal amounts from £50 - £60, that is in total by £10. The 40th value was £50; the median (the 60th value) is 20 values greater. The 60 values increase by £10; therefore 20 values increase by $£(20/60 \times 10 = £3.3$. The median is

£3.3 greater than the 40th value of £50. The median is therefore
£53.3.

The median can also be found graphically by finding the value of
x corresponding to the cumulative frequency of $N/2$ on a cumulative
frequency curve, as shown in Fig. 59.

Note. Do not refer to the median as the "medium". A surprising
number of students do.

PROCEDURE

Finding the median of grouped data by calculation

1. Ensure mathematical limits.
2. Compute cumulative frequencies.
3. Compute rank of median ($N/2$).
4. Locate class containing the median.
5. Deduct cumulative frequency of previous class from rank of
   median.
6. Divide the frequency obtained from step 5 by the frequency
   of the class containing the median and multiply by the class
   interval of the class containing the median.
7. Add the figure obtained in step 6 to the lower limit of the
   class containing the median.

PROCEDURE

Finding the median graphically

1. Draw a cumulative frequency curve.
2. Compute rank of median.
3. Read the value of the variable corresponding to the rank of the
   median (along the cumulative frequency axis).

Note. See Fig. 59 for an illustration of finding the median graphically.

QUARTILES AND OTHER FRACTILES

Just as the median divides the array into two parts, one having half
the values greater than the median, the other having half the values
less than the median, so the *quartiles* divide the array into four parts,
the *deciles* divide it into ten parts and the *percentiles* divide it into
one hundred parts.

The *lower quartile* (denoted by Q_1) will have one-quarter of the values lower and three-quarters of the values higher than the value of the lower quartile. The *upper quartile* (denoted by Q_3) will have three-quarters of the values lower and one-quarter of the values higher than the value of the upper quartile. The second quartile is, of course, the median.

The *first decile* will have one-tenth of the values lower than its value and nine-tenths of the values higher. The *ninth decile* (denoted by D_9) will have nine-tenths of the values lower than its value and one-tenth higher. The *ninety-fifth percentile* (denoted by P_{95}) will have ninety-five per cent of the values lower than its value and five per cent higher.

COMPUTATION OF QUARTILES AND OTHER FRACTILES

The method is exactly the same as for the median. Using the data of Example 9, the calculations are as follows.

Q_1 Rank $= \dfrac{120}{4} = 30$

       Value $= \pounds\left(40 + \left(\dfrac{30-12}{28}\right)10\right)$ $= \pounds46.4$

Q_3 Rank $= \dfrac{3}{4} \times 120 = 90$

       Value $= \pounds\left(50 + \left(\dfrac{90-40}{60}\right)10\right)$ $= \pounds58.3$

D_9 Rank $= \dfrac{9}{10} \times 120 = 108$

       Value $= \pounds\left(60 + \left(\dfrac{108-100}{15}\right)20\right)$ $= \pounds70.7$

P_{95} Rank $= \dfrac{95}{100} \times 120 = 114$

       Value $= \pounds\left(60 + \left(\dfrac{114-100}{15}\right)20\right)$ $= \pounds78.7$

The values of the median, the quartiles and other fractiles can easily be read off a cumulative frequency curve of grouped data. This can be seen from Fig. 59 which shows in graphical form the data given in Example 9.

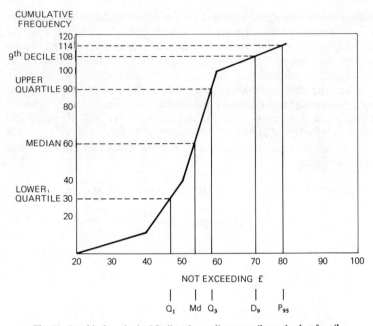

Fig. 59. *Graphical method of finding the median, quartiles and other fractiles.*

THE MODE

In the case of a discrete series it is the most frequently occurring value. To the question "What is the average size of family in the village of X?", the answer expected would be the mode, that is the number of children which occurs more frequently than any other size. The mode is a "typical" value. The reply 2.3 children, an arithmetic mean, usually brings the retort: "How can you have 2.3 children?" However, this would be the required mean if it were being used for further calculations, e.g. forecasting the future population.

In the case of grouped data where the class intervals of the class containing the mode and the classes immediately before and immediately after are all equal, as in Example 10, it is possible to calculate the mode and also to find it graphically.

PROCEDURE

The calculation of the mode

1. Let the frequency in the modal class be denoted by f_1.

2. Let the frequency in the class immediately before be denoted by f_0.
3. Let the frequency in the class immediately after be denoted by f_2.
4. Let the lower limit of the modal class be denoted by l.
5. Let the class interval be denoted by c.
6. Substitute in the following formula:

$$\text{Mode} = l + \left(\frac{f_1 - f_0}{2f_1 - f_0 - f_2} \right) c.$$

Example 10.

x	f	
1.50 - 1.60 m	18	(class immediately before modal class: f_0)
1.60 - 1.70 m	36	(modal class since highest frequency: f_1)
1.70 - 1.80 m	27	(class immediately after modal class: f_2)
1.80 - 1.90 m	9	

$$\text{Mode} = 1.60 + 0.10 \left(\frac{36 - 18}{72 - 18 - 27} \right) = 1.67 \text{ metres.}$$

The mode can also be found graphically as shown in Fig. 60.

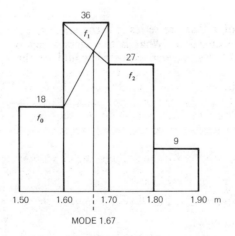

Fig. 60. Finding the mode — graphical method.

It will be seen that the value obtained by the graphical method is identical with that obtained arithmetically.

In the case of a frequency distribution that can be represented by a smoothed histogram or a frequency curve the *mode is the value of highest frequency density* (*see* Fig. 61). Unless the frequency distribution is a mathematical function the mode is often difficult to determine. However, in the case of a moderately skewed distribution the following empirical relationship will give a sufficiently approximate value (*see* Fig. 61):

$$\text{Mode} = \bar{x} - 3(\bar{x} - Md)$$

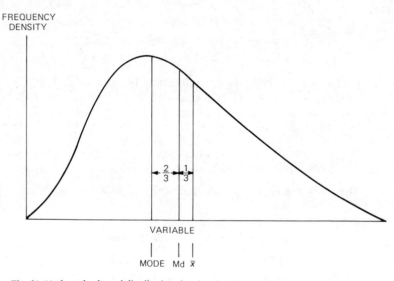

FREQUENCY
DENSITY

$\frac{2}{3}$ ← → $\frac{1}{3}$

VARIABLE

MODE Md $\bar{x}$

Fig. 61. Moderately skewed distribution showing the relationship between mean, median and mode.

BI-MODAL DISTRIBUTION

Sometimes a frequency distribution will have more than one mode. An example of a *bi-modal distribution* is shown in Fig. 62. This means in all probability that the material being dealt with is not homogeneous. An example might be the examination marks for a class of mixed ability. The arithmetic mean mark may well have a low frequency density.

THE GEOMETRIC MEAN

The geometric mean (G.M.) is the nth root of the product of n values.

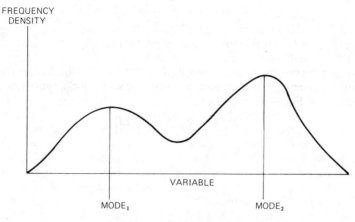

FREQUENCY
DENSITY

VARIABLE

MODE₁   MODE₂

Fig. 62. A bi-modal distribution.

Example 11. What is the geometric mean of 1.05, 1.12, and 1.34?

There are three values. The geometric mean will be the cube root of the product of these three values:

$$\text{G.M.} = \sqrt[3]{1.05 \times 1.12 \times 1.34}$$

Number	*Logarithm*
1.05	0.02119
1.12	0.04922
1.34	0.12710
	3)0.19751
1.1637	0.06584

G.M. = 1.1637

The geometric mean is used to find the average rate of growth or decline.

Example 12. The rate of inflation in three successive years in Barbalonia was 5 per cent, 12 per cent and 34 per cent. What was the average rate of inflation per year?

$$\text{G.M.} = \sqrt[3]{(1.05)(1.12)(1.34)} \quad = \quad 1.1637$$

Average rate of inflation is 16.37 %.

Consider the price of a given basket of commodities. Let it be m at the beginning of the first year. At the end of the first year it will be $1.05\,m$, that is, an increase of 5 per cent. At the end of the second year it will be $(1.05)(1.12)m$, an increase of 12 per cent. At the end of the third year it will be $(1.05)(1.12)(1.34)m$, an increase of 34 per cent over the year. At the end of the third year the price of the basket will be $1.576m$. The average rate of increase is 16.37 per cent. At the end of the first year the price of the basket would have been $1.1637m$. At the end of the second year the price would have been $(1.1637)(1.1637)m$. At the end of the third year the price would have been $(1.1637)(1.1637)(1.1637)m$, an increase of 16.37 per cent for the third time. This would amount to $1.576m$, the same as the varying rates.

Note: It is a gross, but not infrequent error, to take the A.M. instead of the G.M. In Example 12, the A.M. of 5, 12, and 34, i.e. 17, is *incorrect.*

Example 13. During the first five years of its independence the population of Barbalonia increased by 2 per cent per annum and during the next ten by 1 per cent per annum. What was the average annual rate of increase?

Average annual rate of increase:

$$\sqrt[15]{(1.02)^5 (1.01)^{10}} - 1 = 1.0133 - 1 = 0.0133 = 1.33\%.$$

	Arithmetic mean	Harmonic mean	Median	Mode	Geometric mean
Calculation takes into account:	every value	every value	middle value only	value of highest frequency density or the most frequently occurring value	every value

Extreme values have:	greatest effect	least effect	no effect	no effect	less effect than on $\bar{x}$ but more than on the harmonic mean
Applied to the same data:	greater than G.M. and H.M.	smaller than G.M. and $\bar{x}$	between $\bar{x}$ and mode	may be larger or smaller than $\bar{x}$ or median	larger than harmonic mean but smaller than $\bar{x}$
Computation:	possible $\bar{x} = \dfrac{\Sigma fx}{N}$	possible H.M. $= \dfrac{N}{\Sigma\frac{1}{x}}$	not possible: it is a positional average; interpolated value in many cases	not possible in many cases.	possible G.M. $= \sqrt[n]{(x_1)(x_2)(x_3)\ldots(x_n)}$
Open-end classes make:	indeterminate	indeterminate	–	–	indeterminate
Used:	(a) generally (when no reason to use any other) (b) rates of the form x per d and d is constant	rates of the form x per d and x is constant	when representative value required and distribution is skewed	when the most frequently occurring value is required (discrete variable)	when average rate of growth or decline required

Fig. 63. The averages compared.

Note. In a perfectly symmetrical frequency distribution of a continuous variable the arithmetic mean, the mode and the median are equal.

QUESTIONS

1. From the following table showing the wage distribution in a certain factory determine:

    (a) the mean wage,
    (b) the median wage,
    (c) the modal wage,
    (d) the wage limits for the middle 50 per cent of the wage earners,
    (e) the percentage of workers who earned between £70 and £100,
    (f) the percentage who earned more than £60 per week,
    (g) the percentage who earned less than £50 per week.

Weekly wage £	Number of employees
35 - 45	17
45 - 55	23
55 - 65	42
65 - 75	52
75 - 85	78
85 - 95	117
95 - 105	223
105 - 115	82
115 - 125	14
125 - 135	3

2. A man gets three annual rises in salary. At the end of the first year he gets an increase of 4 per cent, at the end of the second an increase of 6 per cent on his salary as it was at the end of the year, and at the end of the third year an increase of 9 per cent on his salary as it was at the end of the third year. What is the average percentage increase?

3. Define a "weighted average".

Dept.	Number of workmen	Total wages
A	432	£34,321
B	517	58,372
C	117	10,498

A cost-of-living bonus is given to each workman amounting to £20. What is the average percentage increase per man for each department and for the total?

4. Compare and contrast the mean, the median and the mode as measures of magnitude for statistical data. Under what circumstances would you use the median or the mode in preference to the mean?
Institute of Chartered Secretaries and Administrators.

5. The values of orders in the order book of a firm have been analysed, and the following grouped frequency table is available.

Outstanding Orders by Value (£00s)		
Value of orders (£00s)		Number of orders outstanding
5 and up to, but less than	7	16
7 ” ”	9	29
9 ” ”	11	56
11 ” ”	12	47

Note that the last group (11 and up to, but less than 12 units) has a different group-width from the others in the table. What is the arithmetic average value of orders, and the total value of all outstanding orders? (Answers should be given correct to the nearest £1.)

Institute of Chartered Secretaries and Administrators.

6. A company which makes and sells a standard article has four machines on which this article can be made. Owing to differences in age and design these machines run at different speeds, as follows:

Machine	Number of minutes required to produce one article
A	2
B	3
C	5
D	6

(a) When all machines are running what is the total number of articles produced per hour?

(b) Over a period of 3 hours, machines B, C and D were run for the first 2 hours, and machines A, B and D only were run for the last hour. What was the average number of articles produced *per hour* over this 3-hour period?

Institute of Cost and Management Accountants.

7. (a) Place the arithmetic mean, median and mode in order of merit as averages for the following frequency distributions, and briefly explain your rankings in each case.

(i) Incomes, taken from a wages survey.

(ii) Ladies' shoe sizes, based on sales data.

(iii) Percentage of defective products, based on batches examined.

(b) The following table shows the number of hours of sunshine recorded during July at Bournpool for the years 1971 - 1979:

Hours of sunshine	Number of days
0 and under 1	1
1 " " 2	2
2 " " 3	4
3 " " 4	11
4 " " 5	24
5 " " 6	35
6 " " 7	43
7 " " 8	49
8 " " 9	54
9 " " 10	31
10 " " 11	15
11 " " 12	10
	279

Calculate

(i) the mean number of hours of sunshine.
(ii) the median number of hours of sunshine.

Association of Certified Accountants.

8. An investment increased in value after one year by 6 per cent. The next year it decreased by 6 per cent. The following year it increased by 9 per cent and at the end of the fourth year it decreased by 9 per cent. What was the average yearly change in value and was it an increase or decrease?

9.

Cinemas in Great Britain

Size of cinema (seats)	Number of cinemas
Up to 250	165
251– 500	897
501– 750	1,043
751–1,000	895
1,001–1,250	589
1,251–1,500	334
1,501–1,750	264
1,751–2,000	206
Over 2,000	204

Using graphical methods estimate the median and quartile size of cinema.

10. The following table shows the frequency distribution of actual headways between buses which were observed at one point on an in-town route with a scheduled frequency of 30 buses per hour. Calculate *(a)* the arithmetic average headway, *(b)* the median headway and *(c)* the two limits—one above and one below the median—between which it is estimated that half the number of headway observations occurred.

Headway (nearest minute)	Number of buses
0	111
1	189
2	117
3	75
4	48
5	30
6	15
7	8
8	3
9	3
10	1

Chartered Institute of Transport.

11. *(a)* Mr. Short went from his home to work by car and his petrol consumption was 8 litres per 100 kilometres. He came back home by exactly the same route and he travelled exactly the same distance as he did going to work, but the journey was now uphill and his petrol consumption suffered accordingly. He needed 9 litres per 100 kilometres. What was the average rate of petrol consumption for the journey there and back each day?

(b) His colleague, Mr. Rounday lived next door. He had a bigger car and his petrol consumption to work was 9 litres per 100 kilometres. He returned home in a leisurely way by a different route and although his journey home was longer, he used the same amount of petrol going home as he did going to work, but his petrol consumption on the way home was only 8 litres per 100 kilometres. What was Mr. Rounday's average rate of petrol consumption for the journey to work and back?

12. In a moderately skewed distribution, the mean was £57 and the median £62. Estimate the value of the mode.

CHAPTER 15

Index Numbers

THE USE OF INDEX NUMBERS

It is useful for many purposes to know how the price level of a group of commodities has changed over time. A price index enables this to be done. It is an average of the changes in the price of the individual items in the group. Similarly a quantity or volume index shows the average changes in the quantitites of the items of a group of commodities. Examples of price indices include the retail price index, the index of wholesale prices, the index of weekly wage rates (wages being the price of labour) and the index of agricultural prices. Quantity indexes include the index of industrial production, and the indices of the volume of exports and imports.

Index numbers are useful for showing trends; they enable comparisons to be made between movements in the levels of prices of different groups of commodities, or between the levels of prices and wages, or between the levels of production and wages, and so on.

Index numbers can also be used to make comparisons over space. It is also possible to construct index numbers to show movements in productivity within a firm. *An index number is a device which shows by its variations the changes in a magnitude which is not capable of accurate measurement in itself or of direct valuation in practice.* It is an indirect measure of a concept, e.g. changes in price level. As such, there are various methods of compiling an index number. Naturally different methods give different results. Since, however, only a series of index numbers has any meaning, provided the same method of compilation is used throughout the series, this is not important.

COMPILATION OF INDEX NUMBERS

It is now proposed to show the usual method of compilation of an index number. Most of the other methods are of academic interest only, and will not be dealt with.

In order to illustrate the method and to keep calculations to a minimum, only three commodities have been chosen. A price index and a quantity index will be computed.

The data are as follows:

| Item | Unit | 3rd August 1975 | | 9th June 1980 | |
		Quantity	Price	Quantity	Price
			p		p
Bread	per loaf	3	8	6	12
Butter	per 250gm	1	45	1½	36
Cheese	per 250gm	½	30	1	36

The quantities are the average weekly purchases of these commodities of an imaginary household for the dates given. The prices are those ruling on the dates given. The price index number for 9th June 1980 will show the price level of these commodities as a percentage of the price level at 3rd August 1975. The quantity index for 9th June 1980 will show the average weekly quantity bought for this date as a percentage of the average weekly quantity bought for 3rd August 1975.

It will subsequently be shown that the price index is 107, which means that prices have risen by 7 per cent between these two dates. It will also be shown that the quantity index is 173, which means that the average weekly quantity bought has increased by 73 per cent.

The date with which comparison is made is known as the base date, in this case 3rd August 1975.

The price index will be computed first. The first step is to calculate the *price relatives*. A price relative is the price at the current date expressed as a percentage of the price at the base date. In the example given the price relatives are as follows:

$$\text{Bread} \left(\frac{12p}{8p}\right)(100) = 150.$$

$$\text{Butter} \left(\frac{36p}{45p}\right)(100) = 80.$$

$$\text{Cheese} \left(\frac{36p}{30p}\right)(100) = 120.$$

To obtain the price index, a weighted average of these price relatives is calculated. The importance of these various commodities, and therefore the weights, will be proportional to the expenditure on each item. The expenditure will be that of the base date. The calculation is shown below.

Item	Expenditure at base date	Price relative	Price relative X weight
	p		
Bread	24	150	3,600
Butter	45	80	3,600
Cheese	15	120	1,800
	84		9,000

$$\text{Price Index} = \frac{9,000}{84} = 107.$$

Index numbers are calculated to the nearest whole number, unless they are to be used for further calculations.

The quantity index is computed in a similar way. It will be a weighted average of quantity relatives, the weights being the same as for a price index. The calculations are as follows:

Item	Quantity relatives	Weight	Quantity relative X weight
Bread	$\left(\dfrac{6 \text{ loaves}}{3 \text{ loaves}}\right)(100)$	24	$(200)(24) = 4,800$
Butter	$\left(\dfrac{375 \text{gm}}{250 \text{gm}}\right)(100)$	45	$(150)(45) = 6,750$
Cheese	$\left(\dfrac{250 \text{gm}}{125 \text{gm}}\right)(100)$	15	$(200)(15) = 3,000$
			14,550

$$\text{Quantity Index} = \frac{14,550}{84} = 173.$$

The above method of calculation is known as the average of ratios method. *The weights are values.*

THE AGGREGATIVE METHOD

This will give the same results as before, provided that base-date quantities and prices are used as weights. The price index will be the value of base-date quantities at current prices, expressed as a percentage of base-date quantities at base-date prices. It expresses the value of a group of commodities at current prices as a percentage of the value of the same group of commodities at base-date prices. The quantity index will be the value of the current quantities at basedate prices expressed as a percentage of the value of the base-date quantities at base-date prices. It expresses the value of current quantities at base-date prices as a percentage of the value of base-date quantities at the same prices.

Using the same data as before, the price index is calculated as follows:

Item	Value at base date	Value of base-date quantities at current prices
	p	p
Bread	24	3 loaves at 12p = 36
Butter	45	1 250gm at 36p = 36
Cheese	15	½ 250 gm at 36p = 18
	84	90

$$\text{Price Index} = \left(\frac{90p}{84p}\right)(100) = 107 \text{ (as before)}.$$

The quantity index is calculated at follows·

Item	Value at base date	Value of current quantities at base-date prices
	p	p
Bread	24	6 loaves at 8p = 48
Butter	45	1½ 250gm at 45p = 67½
Cheese	15	1 250gm at 30p = 30
	84	145½

$$\text{Quantity Index} = \left(\frac{145½p}{84p}\right)(100) = 173 \text{ (as before)}.$$

A theoretical objection to the methods just given is that the price index multiplied by the quantity index will not indicate changes in value. If, however, in calculating the price index, current quantities had been taken instead of base-date quantities, the quantity index being calculated with base-date prices, the product would have shown the changes in value. There is, however, the very serious practical difficulty that when compiling price indices, it is often not possible to obtain current quantities. In practice, the theoretical objection is not important.

THE CHAIN BASE METHOD

Circumstances may arise in which it is possible to compare the prices of one year with the next, but it is not possible to compare prices when separated by a large number of years. If new commodities are replacing old ones, or the weights are rapidly changing, the methods just described will not be possible. In this case, it may be useful to compile indices with the previous year as base and then "chain" them so as to get a series referring back to a base year. Movements between successive years will be as accurate as those compiled by the methods just described, but for movements between years far apart, the series may be very inaccurate.

Example.
        Price index for 1969 (1968 = 100) = 110.
        Price index for 1970 (1969 = 100) = 105.
        Price index for 1971 (1970 = 100) = 125.

Chain indices are obtained as follows:

Price index for 1970 (1968 = 100) $= \dfrac{(110)(105)}{100} = 115.5$

Price index for 1971 (1968 = 100) $= \dfrac{(110)(105)(125)}{(100)(100)} = 144.$

The series (1968 = 100) is 100; 110; 116; 144.

PROBLEMS OF INDEX NUMBER CONSTRUCTION

The first essential point to be considered is the object for which the index number is to be constructed. What is it to measure and why? Suppose, for example, it is required to measure the amount of production to show the trend of economic activity. Suppose, further, this is required monthly. The data which can be made available will determine the scope of the index. Since it is to be monthly, agricultural production will have to be omitted. It will also be difficult to include many forms of production. The problem of the scope of the index is bound up with the purpose of the index and the data available. The data avilable, or rather the lack of it, may necessitate the modification of the purpose. Instead of an index of production as a whole being sought, an index of industrial production might now become the aim.

The next problem is the selection of items. Consider the compilation of a price index. Usually, it is not practicable, either from the point of view of cost or time, even if it is possible, to measure changes in the prices of all the relevant commodities. Hence a selection must be made. These must be selected so that movements in the prices of those chosen will be representative of the movements of prices of all the relevant commodities.

The choice of the base date or period presents some difficulties. However, there is a general misconception that a "normal" year or date must be chosen, otherwise the weights soon become incorrect. In the case of a quantity index the base date or period must only be such that the prices of the various commodities are reasonably related, and in the case of a price index, a base date or period when the quantities of the commodities are reasonably related should be chosen.

The choice of weights provides another set of problems. Where fewer items are chosen as indicators than all those relevant, the weight must be that relating to the whole of those relevant. Thus, in the case of the index of retail prices, only five items are taken to represent floor coverings, but the weight is in respect of floor coverings as a whole. In the case of the index of production, the weights used are the values of net outputs of the various industries. Sometimes difficulty arises because the value of the net output of an industry has to be apportioned over a number of products.

The form of average is usually dictated by practical considerations. Thus, despite the theoretical objections to the methods described earlier in the chapter, they are the most usual.

LIMITATIONS OF INDEX NUMBERS

The index is usually based on a sample, hence sampling errors are introduced. It is not possible to take into account all changes in quality or product. Comparisons over long periods are not reliable. Different methods of compilation give different results, but, unless there are rapid changes in conditions, the trends generally agree.

INDEX NUMBERS FORMULAE

p_0 represents the price at the base date.
p_1 represents the price at the current date.
q_0 represents the quantity at the base date.
q_1 represents the quantity at the current date.

(a) Price Index.

Average of ratios $\qquad \dfrac{\Sigma \dfrac{p_1}{p_0}(p_0 q_0)}{\Sigma p_0 q_0}$.

Aggregative method $\dfrac{\Sigma p_1 q_0}{\Sigma p_0 q_0}$.

(b) Quantity Index.

Average of ratios $\qquad \dfrac{\Sigma \dfrac{q_1}{q_0}(p_0 q_0)}{\Sigma p_0 q_0}$

Aggregative method $\dfrac{\Sigma q_1 p_0}{\Sigma q_0 p_0}$.

Both these index numbers are *Laspeyres* indices; they have base-period quantities and base-date prices for weights. Index numbers which have current-period quantities and current-date prices are called *Paasche* indices.

QUESTIONS

1. It has been stated that the technique of index number construction involves four major factors:

    (a) choice of items,
    (b) base period,
    (c) form of average,
    (d) weighting system.

Do you agree with this view? If so, explain these four factors and discuss the problems to which they give rise. If you do not agree,

give your views on the main problems involved in index construction.
Chartered Institute of Transport.

2. In the manufacture of a particular product, two commodities, X and Y, are considered to be equally important in 1971. Their prices in June 1971 and 1972 were as follows:

	1971	1972
X £ per kg . .	8	12
Y pence per metre . .	30	20

Show the unweighted price relatives for each commodity using 1971 and 1972 as bases, and comment on the results.

How would a combined index for X and Y be worked out?

3.   *Barbalonia Imports of Dutiable Beverages*

	1960		1979	
	Quantities	*Value (£)*	*Quantities*	*Value (£)*
Beer and Ale (barrel)	58,577	158,650	448	4,057
Cocoa, raw (kg)	51,670,320	1,335,107	246,623,216	8,943,025
Coffee (tonnes)	765,561	2,024,648	1,066,046	5,988,812
Tea (kg)	21,190,064,	9,904,085	494,353,466	33,050,853
Wine (litres)	13,103,304	4,214,878	25,252,387	18,167,077

Compute *(a)* a volume index for 1979, using 1960 as the base date; *(b)* a price index for 1960, using 1979 as base date.

(*Note:* Average price for each commodity is found by dividing values by quantitites. Hence price relatives can be found.)

4. The following table indicates the price per tonne, and value of sales for a raw material.

Grade of material	*1968*		*1974*	
	Price (£)	*Sales (£ mn.)*	*Price (£)*	*Sales (£ mn.)*
A	6.50	13	8.75	26.25
B	9.00	81	14.00	140.00
C	8.50	34	10.00	50.00

Calculate a Laspeyres Index for 1974 (all grades) taking 1968 as 100.   *Institute of Chartered Secretaries and Administrators.*

5. A steel stockist notices that prices and values of sales for the main units of steel supply were:

Type of steel items	1969 Price per tonne (£)	Sales (£ mn.)	1974 Price per tonne (£)	Sales (£ mn.)
Ingots	162	324	200	600
Steel bars	188	564	190	760
Steel strip	220	880	275	1,100

Calculate a Paasche Index number of 1974 prices, taking 1969 = 100.

Institute of Chartered Secretaries and Administrators.

CHAPTER 16

Dispersion and Skewness

Frequency distributions may well have the same mean and yet be very different. Figure 64 shows the frequency curves of three distributions having the same mean.

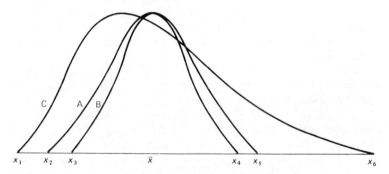

Fig. 64. Frequency distributions with same mean but varying dispersions.

The values of distribution B vary from x_3 to x_4; they are fairly near the mean value. For distribution A the values are more widely *dispersed,* from x_2 to x_5 and in distribution C the variation is even greater, x_1 to x_6. Measures showing the extent of the variation of individual values around the mean are *measures of dispersion.* It will also be seen that the distributions A and B are symmetrical whereas C is *skewed* (i.e. asymmetrical). The shape of the distribution can be shown by *measures of skewness.*

Example 1. The following three distributions all have the same mean value, namely 10 tonnes. However, the individual values are

closest to the mean in the case of Q and are spread most widely in
the case of R.

> P - 2, 6, 14 and 18 tonnes.
> Q - 8, 9, 11 and 12 tonnes.
> R - 1, 8, 11 and 20 tonnes.

MEASURES OF DISPERSION

The simplest method of indicating the amount of dispersion is to
take the difference between the lowest and highest values. This is
known as the *range*. For Example 1, this would be in the case of
distribution P, 16 tonnes, in the case of distribution Q, 4 tonnes,
and in the case of distribution R, 19 tonnes.

The range, however, ignores all values except the extreme values
and is of little practical importance apart from its use in
quality control (*see* Chapter 23).

The most obvious method would appear to be calculating the
arithmetic mean of the deviations of each value from the mean of
the distribution. But the sum of the deviations is zero. However, if
the signs of the deviations are ignored (an illogical procedure), this
will give the *mean deviation*. For Example 1 this would be for
distribution P, 6 tonnes, for distribution Q, 1.5 tonnes, and for
distribution R, 5.5 tonnes.

There is, however, an average of the deviations from the mean of
the distribution that does not ignore the sign of the deviation—the
quadratic average. It is the *root mean square*.

The *standard deviation*, by far the most important measure of
dispersion, is the quadratic mean of the deviations of all the values
of the distribution from the arithmetic mean of the distribution.

The standard deviation (denoted by S.D. or σ, the Greek letter
sigma) may be given as $\sqrt{(\Sigma d^2/N)}$. The alternative formula
$\sqrt{(\Sigma x^2/N) - \bar{x}^2}$, which gives the same result, is used in practice.

Example 2. The deliveries of coal in four successive quarters were
8, 9, 11 and 12 tonnes. What was the mean quarterly amount
delivered and what was the standard deviation?

x tonnes	d $(x-x)$	d^2	x^2
8	−2	4	64
9	−1	1	81
11	1	1	121
12	2	4	144
$\Sigma x = 40$		$\Sigma d^2 = 10$	$\Sigma x^2 = 410$

$$\bar{x} = \frac{\Sigma x}{N} = \frac{40}{4} = 10 \text{ tonnes.}$$

S.D. $= \sqrt{(\Sigma d^2 / N)} = \sqrt{(10/4)} = 1.58$ tonnes.

Using alternative formula:

S.D. $= \sqrt{(\Sigma \bar{x}^2 / N) - \bar{x}^2} = \sqrt{(410/4) - 10^2} = 1.58$ tonnes.

Note. The standard deviation like the mean will *not* be an abstract quantity. It will show kilometres, hours, tonnes, or some other unit.

Another measure of dispersion is the *quartile deviation*, also known as the *semi-interquartiles range*. Like the range only two values are used to calculate it, the upper and lower quartiles. The formula is:

$$\text{Q.D.} = \frac{1}{2}(Q_3 - Q_1)$$

Example 3. Find the quartile deviation of the wage distribution given in Example 4.

$$\text{Q.D.} = \frac{1}{2}(58.3 - 46.4) = \pounds 5.95.$$

THE STANDARD DEVIATION OF GROUPED DATA

In Example 2 each value had a frequency of one. In the case of grouped data, values have a frequency of more than one; the deviations have, therefore, to be multiplied by their respective frequencies. The formula for standard deviation becomes:

$$\text{S.D.} = \sqrt{\frac{\Sigma f d^2}{N}} = \sqrt{\frac{\Sigma f x^2}{N} - \bar{x}^2}.$$

Example 4.

x Wages (£)	f	fx	fx^2	$x + 1$	$f(x + 1)$	$f(x + 1)^2$
20 - 40	12	360	10,800	31	372	11,532
40 - 50	28	1,260	56,700	46	1,288	59,248
50 - 60	60	3,300	181,500	56	3,360	188,160
60 - 80	15	1,050	73,500	71	1,065	75,615
80 - 100	5	450	40,500	91	455	41,405
	120	6,420	363,000		6,540	375,960

$$N = 120 \qquad \Sigma fx = 6{,}420 \qquad \Sigma fx^2 = 363{,}000$$
$$\Sigma f(x + 1) = 6{,}540 \qquad \Sigma f(x + 1)^2 = 375{,}960$$

$$\bar{x} = \frac{\Sigma fx}{N} = \frac{6{,}420}{120} = £53.50$$

$$\text{S.D.} = \sqrt{\frac{\Sigma fx^2}{N} - \bar{x}^2} = \sqrt{\frac{363{,}000}{120} - 53.5^2} = £12.76$$

Charlier's check: Verify that

$$\Sigma f(x + 1)^2 = \Sigma fx^2 + 2\Sigma fx + N$$
$$375{,}960 = 363{,}000 + 12{,}840 + 120.$$

(Charlier's check for the mean is given in Chapter 14.)

Note. Do *not* square the *fx* column to obtain the *fx²* column. This would square the frequency as well as the variable. Multiply the *fx* column by the *x* column.

Similarly, do *not* square the *f(x + 1)* column, but multiply the *f(x + 1)* column by the *x + 1* column to obtain the *f(x + 1)²* column.

PROCEDURE

Computing the standard deviation and arithmetic mean of grouped data

1. Ensure mathematical limits.
2. Make reasonable assumptions for any open-ended classes.
3. Calculate mid-points of classes.
4. *(a)* Complete columns *fx* and *fx²*.
   (b) If using Charlier's check, complete columns *x + 1, f(x + 1)* and *f(x + 1)²*.
5. *(a)* Sum columns *f, fx* and *fx²*.
   (b) If using Charlier's check, sum columns *f(x + 1)* and *f(x + 1)²*.
6. *(a)* Substitute for *N* and *Σfx* in formula $\bar{x} = \dfrac{\Sigma fx}{N}$ and then substitute for *N*, *Σfx²* and *x̄* in formula

$$\text{S.D.} = \sqrt{\frac{\Sigma fx^2}{N} - \bar{x}^2} = \sqrt{\frac{\Sigma fx^2}{N} - \left(\frac{\Sigma fx}{N}\right)^2}$$

   (b) If using Charlier's check, verify:

(i) $N + \Sigma fx = \Sigma f(x + 1)$ (the arithmetic check for mean);
(ii) $N + 2\Sigma fx + \Sigma fx^2 = \Sigma f(x + 1)^2$ (the arithmetic check for the standard deviation).

RELATIVE MEASURES OF DISPERSION

The measures of dispersion so far dealt with (the mean deviation, the quartile deviation and the standard deviation) are all expressed as so many metres, tonnes, litres, etc. When it is required to compare the variability of the data of two or more distributions, then relative measures of dispersion must be used, that is, measures that are independent of the units used to measure the variable.

Suppose it is required to know whether the height (measured in metres) of a certain group of men varies more or less than their weight (measured in kilograms). Again, suppose it is required to know whether the incomes of shop assistants (measured in £s) vary more or less than the incomes of chartered accountants (also measured in £s). In both these cases relative measures of dispersion must be used. These are abstract quantities, i.e. not "concrete" measures in kilograms, £s etc.

The relative measures of dispersion are:

(a) coefficient of mean deviation $= \dfrac{\text{mean deviation}}{\text{mean}}$.

(b) coefficient of quartile deviation $= \dfrac{Q_3 - Q_1}{Q_3 + Q_1}$.

(c) coefficient of standard deviation $= \dfrac{\text{S.D.}}{\text{mean}}$.

COEFFICIENT OF VARIATION

The principal measure of relative dispersion is the coefficient of variation which expresses the standard deviation as a percentage of the mean.

Example 5. What is the coefficient of variation of the wage distribution of Example 4?

$$\text{C.V.} = \frac{12.76}{53.50} = 23.85\%.$$

Note. When comparing the variation of two or more frequency distributions the coefficient of variation must be used.

PROCEDURE

Comparing the variation of two or more frequency distributions

1. Compute the mean and standard deviation of each distribution.
2. Compute the coefficient of variation for each distribution
   (C.V. = $(\frac{S.D.}{mean} \times 100)\%$).
3. Compare the coefficients of variation.

It is also possible to compare the variation of frequency distributions by graphical means. Refer to the paragraphs on the Lorenz curve in Chapter 11.

SKEWNESS

Figures 65 and 66 both show a frequency curve AB of a symmetrical distribution. The arithmetic mean (the "balancing" value), the median (which divides the area under the frequency curve into two equal parts) and the mode (the value below the peak of the curve) are all equal.

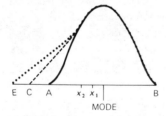

Fig. 65. *Negatively skewed distribution.*   Fig. 66. *Positively skewed distribution.*

Consider now Fig. 65. If more values less than the mean are added to the existing symmetrical distribution AB, the new mean, say x_1, will become less than the previous mean and therefore less than the mode. The new distribution denoted by the curve CB is now skewed. If even more values less than the mean are added, the new mean will become still less, say, x_2 and the distribution denoted by EB will become even more skewed; the increase in skewness being reflected by the increasing difference between the value of the mode and the mean.

Consider now Fig. 66. In this case adding values above the mean has increased the mean to x_3 for distribution AD and to x_4 for distribution AF. Again it is seen that increasing skewness results in an increasing difference between the mode and the mean.

The difference between the median and the mean will also increase with increasing skewness but not to the same extent (*see* Fig. 61 in Chapter 14).

POSITIVE AND NEGATIVE SKEWNESS

When the mean is greater than the mode, the skewness is positive (Fig. 66). There are more values below the mean than above. An example is the distribution of incomes.

When the mean is less than the mode, skewness is negative (Fig. 65). There are more values above the mean than below. An example is the distribution of the age of death.

THE MEASUREMENT OF SKEWNESS

The fact that the mode and mean differ in a skewed distribution is the basis of the *Pearsonian* measure of skewness; its formula is

$$Sk. = \frac{mean - mode}{S.D.}.$$

The Pearsonian measure of skewness varies between limits of +3 and −3 but values greater than +1 or less than −1 are infrequent.

Example 6. Compute the Pearsonian measure of skewness for the wage distribution of Example 4.

For this distribution the mode is badly defined. The alternative formula $Sk. = \dfrac{3 \ (mean - median)}{S.D.}$ is therefore used (*see* Fig. 61 in Chapter 14):

$$Sk. = \frac{3(53.50 - 53.33)}{12.76} = 0.04$$

PROCEDURE

Computing the Pearsonian measure of skewness

1. Compute the mean and the standard deviation.
2. Compute the mode, or if this is not possible the median.
3. Substitute in relevant formula, either

$$\frac{(mean - mode)}{S.D.} \qquad \text{or} \qquad \frac{3(mean - median)}{S.D.}$$

There is also a *quartile measure of skewness* the formula for which is

$$\frac{Q_1 + Q_3 - 2 \text{ (median)}}{\frac{1}{2}(Q_3 - Q_1)}.$$

This can vary between +2 and −2.

QUESTIONS

1. From the following figures say which class of grocers vary more in size assuming that the value of sales is the measure of size.

Value of sales	Grocers with meat	Grocers with off-licence
Under £1,000	29	128
£1,000–£2,499	189	766
£2,500–£4,999	626	2,130
£5,000–£9,999	673	2,437
£10,000–£24,999	761	2,255
£25,000–£49,999	286	772
£50,000–£99,999	74	170
£100,000–£250,000 £250,000 and over	7	45
	2,645	8,703

(*Note.* It is necessary to compare the C.V.)

Institute of Statisticians.

2.

Age Distribution of Women, Barbalonia, 1979

Years of age	Millions
Under 10	3.7
10 and under 20	3.3
20 ” 30	3.7
30 ” 40	3.9
40 ” 50	3.7
50 ” 60	3.1
60 ” 70	2.4
70 ” 80	1.4
80 and over	0.4

Calculate the standard deviation and coefficient of dispersion for the above distribution. Mention any other method of calculating dispersion with which you are acquainted.

3. (a) Distinguish between absolute and relative dispersion.

(b) In a final examination in statistics the mean mark of a group of 150 students was 78 and the standard deviation 8; in accountancy the mean mark of the group was 73 and the standard deviation was 7.6.

In which subject was the relative dispersion greater?

Association of Certified Accountants.

4.

Football Pools

Amount of stake		Number of individuals
6p and under 12p		35
12p " 24p		105
24p " 36p		76
36p " 48p		52
48p " 60p		35
60p " 72p		26
72p and over		11
		340

Calculate the standard deviation and the arithmetic mean of the given distribution. Assuming the mode is about 18p, estimate the skewness of the distribution.

5. Construct a histogram showing the distribution of football stakes given in Question 4. In your diagram, insert the median and the upper and lower quartiles. Find the ratio of the interquartile range to the total range. Take the last group to be 72p and under 84p.

6. 100 containers of agricultural foodstuffs were checked for weight (each contained nominal weight 50 kilograms), and the following table of results was obtained:

Container weight (kilograms)	Number of containers
49 and up to, but less than 50	11
50 " " " " " " 51	17
51 " " " " " " 52	43
52 " " " " " " 53	29
Total number of containers	100

Calculate the arithmetic average weight and the standard deviation of weights.

(*Note.* Use all relevant checks with your working.)

London Chamber of Commerce and Industry.

7. Define the following measures of dispersion:

(*a*) standard deviation,

(*b*) quartile deviation.

Calculate the standard deviation of the following distribution related to the arrival times of aircraft at U.K. airports. What does it reveal?

No. of aircraft	Minutes late on arrival
14	6
37	4
4	7
30	5
15	3

Chartered Institute of Transport.

8. (*a*) Explain and illustrate graphically the relationship between the mean, median and mode in (*i*) a positively skewed frequency distribution and (*ii*) a negatively skewed distribution.

(*b*) Calculate the coefficient of skewness for a frequency distribution with the following values: mean = 10; median = 11; standard deviation = 5. What does your answer tell you about this frequency distribution? *Association of Certified Accountants.*

CHAPTER 17

Correlation

THE MEANING OF CORRELATION

If two quantities vary in such a way that movements in one are accompanied by movements in the other, these quantities are correlated. Many examples will come to mind. An increase in the amount of rain will be accompanied by an increase in the sales of umbrellas; older men usually have older wives than younger men; an increase in rainfall will up to a point be accompanied by an increase in the output of wheat per acre; and an increase in the issue of television licences is accompanied by a decrease in the number of cinema admissions. In each of the examples given, movements in one variable are accompanied by movements in the other. In the first three cases the movements were in the same direction, an increase of one variable was associated with an increase in the other variable. The correlation was positive. In the last case the movement of one variable was associated with a movement in the other variable in the opposite direction, an increase of one being accompanied by a decrease in the other. The correlation was negative.

It is important to realise that although two quantities move in sympathy, changes in one do not necessarily cause changes in the other. Correlation is *not* synonymous with causation. The changes may both be due to a common cause. Thus, there is a considerable degree of correlation between deaths due to cerebral thrombosis and the number of old-age pensions. It cannot be said that old-age pensions cause cerebral thrombosis. The increase in both is due to a common cause, namely that people are living longer. There may be no connection whatsoever between the two quantities, apart from the fact that they just happen to move in sympathy. An example of such "nonsense correlations", as they are termed, is the considerable degree of correlation stated to exist between the number of births in the U.K. and the stork population of Scandinavia! Whether the movements indicate causation or not must be decided on other evidence than the degree of correlation.

SCATTER DIAGRAMS

A useful method of investigating if there is any correlation between two variables is to draw a scatter diagram. In respect of each observation, the value of one variable will be measured along the *y*-axis, and the corresponding value of the other variable will be plotted along the *x*-axis.

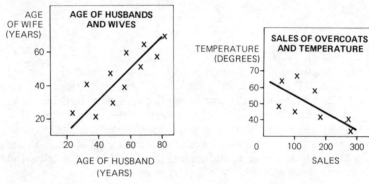

Fig. 67. Linear correlation.   Fig. 68. Inverse correlation.

EXAMPLES OF SCATTER DIAGRAMS

Figure 67 shows in respect of each married couple the age of the husband against the age of the wife. It will be seen that in most cases, but not in all, the older the husband the older the wife. It will also be noted that for two husbands of the same age, the two wives may have different ages. However, there is a distinct tendency for the ages of wives and husbands to move in the same direction. The points plotted cluster along a straight line, a kind of average line. Since the points are near this line, there is a considerable degree of correlation. Since the line of "best fit" is a straight line, the correlation is linear. Further, since the movements of the variables are in the same direction, the correlation is positive. When the points are near the line, it is possible to draw it by inspection. A good method is to stretch a piece of thread and to try out the best position for the line. When, however, the points are too scattered to judge, recourse must be had to mathematical methods. Figure 68 shows inverse linear correlation. Figure 69 shows an example of a scatter diagram where a curve gives the best indication of the relationship between the two variables. This is an example of curvilinear correlation. This kind of correlation is outside the scope of this book. Finally, Fig. 70 gives an example where there is no apparent correlation. The points do not appear to move in any direction.

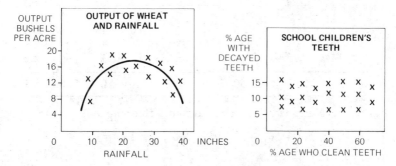

Fig. 69. Curvilinear correlation.   Fig. 70. Zero correlation.

THE COEFFICIENT OF CORRELATION

This is the measurement of the degree of correlation between variables. It will vary between +1 and −1. If there is perfect correlation, the coefficient will be +1 in the case of positive correlation (movement in the same direction) and −1 if there is negative, or inverse correlation. In the case of perfect correlation, if one quantity is known, the other can be calculated with certainty. Thus with a given voltage it is always possible to calculate the amount of electrical current when the resistance is known. There would be no need to use statistical methods. In business, economic and sociological inquiries the coefficient of correlation is most likely to be less than 1. Given the value of one variable, the probable value and not the exact value of the other variable can be found.

COMPUTATION OF THE COEFFICIENT OF CORRELATION
The following formula only applies to linear correlation:

$$r = \frac{\Sigma xy}{\sqrt{(\Sigma x^2 \, \Sigma y^2)}}$$

where r is the coefficient of correlation, x is the deviation from the mean of one variable and y is the deviation from the mean of the corresponding variable.

Example 1. Find the coefficient of correlation between the number of weeks growth and the height in inches of a certain plant.

Plant	X (weeks)	Y (cm)	x	y	x^2	y^2	xy
1	2	10	−2	−5	4	25	10
2	3	5	−1	−10	1	100	10
3	5	20	+1	+5	1	25	5
4	6	25	+2	+10	4	100	20
	16	60			10	250	45

Note. The data relate to four different plants.

$$\Sigma x^2 = 10; \qquad \Sigma y^2 = 250; \qquad \Sigma xy = 45.$$

The mean of the X variable is 16/4 = 4.

The mean of the Y variable is 60/4 = 15.

The xs and ys are deviation from these means.
The column headed xy lists the products of these deviations.

Substituting these values in the above formula:

$$r = \frac{45}{\sqrt{(250)(10)}} = \frac{45}{50} = 0.9.$$

In the example given, the means were whole numbers. In practice, they would probably not be so. To avoid tedious calculations deviations are taken from assumed means and a more complicated formula used. The assumed means are usually taken as zero, so that the deviations from the assumed means are the original data. A table of squares is given in Appendix B to enable the squares of the deviations to be obtained without calculation.

The necessary formula when deviations from assumed averages are taken is:

$$r = \frac{\dfrac{\Sigma XY}{n} - \left(\dfrac{\Sigma X}{n}\right)\left(\dfrac{\Sigma Y}{n}\right)}{\sqrt{\left\{\dfrac{\Sigma X^2}{n} - \left(\dfrac{\Sigma X}{n}\right)^2\right\}\left\{\dfrac{\Sigma Y^2}{n} - \left(\dfrac{\Sigma Y}{n}\right)^2\right\}}}$$

This formula is not so formidable as it seems. The denominator is merely the product of the standard deviations of the two variables.

Using the data of the previous example, the coefficient of correlation will now the re-computed, using the formula, where the deviations are taken from assumed means. The assumed means are taken as zero for both variables.

Example 2.

X	Y	X^2	Y^2	XY
2	10	4	100	20
3	5	9	25	15
5	20	25	400	100
6	25	36	625	150
16	60	74	1,150	285

$\Sigma X = 16; \Sigma Y = 60; \Sigma X^2 = 74; \Sigma Y^2 = 1,150; \Sigma XY = 285; n = 4.$

Substituting these values in the formula

$$r = \frac{\dfrac{285}{4} - \left(\dfrac{16}{4}\right)\left(\dfrac{60}{4}\right)}{\sqrt{\left\{\dfrac{74}{4} - \left(\dfrac{16}{4}\right)^2\right\}\left\{\dfrac{1150}{4} - \left(\dfrac{60}{4}\right)^2\right\}}} = 0.9 \text{ (as before)}.$$

As many of the following operations can be carried out as is useful on either or both of the variables, and the resulting adjusted variables will have the same coefficient of correlation as the original variables.

(a) Add or subtract any amount to each of the values of the X variable. The same amount must be added to, or subtracted from, each item.

(b) Add or subtract any amount to each of the values of the Y variable, not necessarily the same as for the X variable. The same amount must be added to, or subtracted from, each item.

(c) Divide or multiply each value of the X variable by any amount. Each item must be multiplied or divided by the same amount.

(d) Divide or multiply each value of the Y variable by any amount, not necessarily by the same amount as for the X variable. Each item must be multiplied or divided by the same amount.

If the pairs of values are plotted on a scatter diagram, the position of the points indicates the amount of correlation. The effect of the above operations will not affect their position; the adding or subtracting alters the position of the axes, not the relative position of the points, the multiplying or dividing alters the scale without altering the relative position of the points. Hence the coefficient of correlation is unchanged.

Example 3.

X	Y	X − 2	$\frac{Y}{5}$	X − 2	$\frac{Y}{5} - 4$
2	10	0	2	0	−2
3	5	1	1	1	−3
5	20	3	4	3	0
6	25	4	5	4	1

Starting with the same figures as before, 2 was subtracted from each of the values of the X variable, and the Y variable divided throughout by 5. Then 4 was subtracted from each of the values of the adjusted Y variable.

Example 4.

X (adjusted)	Y (adjusted)	X^2	Y^2	XY
0	−2	0	4	0
1	−3	1	9	−3
3	0	9	0	0
4	1	16	1	4
8	−4	26	14	1

$\Sigma X = 8; \Sigma Y = -4; \Sigma X^2 = 26; \Sigma Y^2 = 14; \Sigma XY = 1.$

$$r = \frac{\frac{1}{4} - \left(\frac{8}{4}\right)\left(\frac{-4}{4}\right)}{\sqrt{\left\{\frac{26}{4} - \left(\frac{8}{4}\right)^2\right\}\left\{\frac{14}{4} - \left(\frac{-4}{4}\right)^2\right\}}} = 0.9 \text{ (as before)}.$$

PROCEDURE

Computing the coefficient of correlation

1. Compute X^2, Y^2, XY.
2. Compute ΣX, ΣY, ΣX^2, ΣY^2, ΣXY.
3. Compute the co-variance of X and Y as follows:

$$\frac{\Sigma XY}{n} - \left(\frac{\Sigma X}{n}\right)\left(\frac{\Sigma Y}{n}\right).$$

4. Compute the standard deviations of X and Y.
5. Compute the coefficient of correlation as follows:

$$\frac{\text{COV }(XY)}{\sigma_X \sigma_Y}.$$

TEST OF SIGNIFICANCE OF THE COEFFICIENT OF CORRELATION

When the number of observations is small, it is not possible to place any reliance upon the value of "r".

If the value of "r" exceeds $\dfrac{3}{\sqrt{N-1}}$, where N is the number of observations, it can be assumed that the value of "r" is significant, that it does measure the amount of correlation.

THE COEFFICIENT OF CORRELATION FOR GROUPED DATA

From the figures below, calculate a coefficient of correlation between the age of the workers and the number of days lost through illness in a given year.

	1 or 2	3 or 4	5 or 6	7 or 8	9 or 10	over 10
Under 30	·	2	1	·	·	·
30 - 39	1	2	3	1	·	1
40 - 49	·	1	4	2	3	·
50 - 59	1	1	·	1	·	·
over 60	·	·	·	·	·	1

Illness of Employees in P.Q. Co.

Days lost \ Age	Under 30	30-39	40-49	50-59	60 and over	f	y	fy	fy^2
1 or 2		2 **1*** 2		-2 **1** -2		2	-2	-4	8
3 or 4	2 **2** 4	1 **2** 2	**1**	-1 **1** -1		6	-1	-6	6
5 or 6	**1**	**3**	**4**			8	0	0	0
7 or 8		-1 **1** -1	**2**	1 **1** 1		4	+1	4	4
9 or 10			**3**			3	+2	6	12
over 10		-3 **1** -3		6 **1** 6		2	+3	6	18
f	3	8	10	3	1	25		6	48
x	-2	-1	0	+1	+2			Σfy	Σfy^2
fx	-6	-8		3	2	-9	Σfx		
fx^2	12	8		3	4	27	Σfx^2		
fxy	4	0		-2	6	8	Σfxy		

* The bold figures are frequencies. The figures on the top left of these frequencies are the product of x and y (the deviations from assumed means in working units). The figures on bottom right are the product of the frequency and xy.

$$r = \frac{\dfrac{\Sigma fxy}{n} - \left(\dfrac{\Sigma fx}{n}\right)\left(\dfrac{\Sigma fy}{n}\right)}{\sigma_x \sigma_y} \qquad r = \frac{\dfrac{8}{25} - \left(\dfrac{-9}{25}\right)\left(\dfrac{6}{25}\right)}{\sqrt{\left(\dfrac{27}{25} - \left(\dfrac{-9}{25}\right)^2\right)\left(\dfrac{48}{25} - \left(\dfrac{6}{25}\right)^2\right)}} = 0.3.$$

RANK CORRELATION

Instead of finding the coefficient of correlation between the actual values of two variables it is often sufficiently accurate to rank the variables in order of size; for example, instead of giving actual marks of (say) 6 pupils, give their positions, that is ranks, thus:

Pupil	A	B	C	D	E	F	
French	4	1	2	5	6	3	
Mathematics	1	3	2	4	5	6	
d (difference in ranks)	3	−2	0	1	1	−3	
d^2	9	4	0	1	1	9	$\Sigma d^2 = 24$

$$R = 1 - \frac{6\Sigma d^2}{n(n^2 - 1)} = 1 - \frac{(6)(24)}{(6)(35)} = 0.3.$$

QUESTIONS

1. The following expenditure per child on clothing was obtained from an investigation carried out in York in 1950. Calculate the mean expenditure and the standard deviation.

Values to the nearest £

35 28 17 16 33 14 22 35 18 15 18 19 25 25 17 24
18 16 22 29 30 31 19 26 14 54 31 17 51

Source: *Poverty and the Welfare State.*
London Chamber of Commerce and Industry.

2. Draw a scatter diagram using the data given in Question 1 and the following expenditure on mothers' clothing. Hence, or otherwise, discuss the correlation between the two series of expenditure.

Expenditure on mothers' clothing
Values to the nearest £

46 25 23 47 32 30 11 35 15 28 14 22 51 28 33 43
26 16 31 30 28 35 22 29 45 27 19 23 41

Source: *Poverty and the Welfare State.*
London Chamber of Commerce and Industry.

3.

Merchandise: average value 1970 = 100
Average value

	1969	1970	1971	1972	1973	1974	1975
Imports	90	100	102	115	150	149	132
Exports	92	100	103	108	125	134	130

Discuss the amount of correlation between the average values of exports and imports by means of the scatter diagram and the correlation coefficient.

4. Show the following two series graphically so that any association between them is clearly brought out. If they are related, give an estimate of the relationship.

Series A	92	95	98	103	102	98	106	103	93	96
Series B	96	88	94	93	98	99	101	103	100	97

Series A	94	88	89	99	100	107	94	98	92	102
Series B	98	94	94	99	104	105	93	96	94	100

Chartered Institute of Transport.

5.

American Railroads

	Labour cost % of gross revenue	Average journey per passenger (miles)
1936	42.9	45.7
1937	44.8	49.6
1938	46.5	47.8
1939	44.1	50.3
1940	43.2	52.5
1941	41.1	60.5
1942	37.8	80.4
1943	36.9	99.6
1944	38.7	105.0
1945	41.4	102.9
1946	52.1	81.9
1947	47.6	65.3
1948	46.9	64.1
1949	48.9	63.3
1950	46.2	65.3
1951	48.2	71.5
1952	47.9	72.4
1953	47.5	69.3

How would you establish the existence of any link between the two series given above? How would you measure the degree of linkage and how can you test the significance of your results?

6.

	1966–7	1967–8	1968–9	1969–70
	(millions)			
Number of inland trunk calls	205	217	226	235
Number of inland telegrams	53	47	43	42

Draw the scatter diagram and calculate a coefficient or correlation. Explain carefully, by means of the scatter diagram (or otherwise) the meaning of correlation. What does the correlation coefficient measure?

7. The following table shows the trend of cinema admissions and the growth of television licenses in the Sutton Coldfield area during 1950–52.

		Cinema admissions (thousands)	Television licences per 1,000 population
1950	3rd quarter	10,025	24
	4th "	9,924	37
1951	1st "	9,814	52
	2nd "	9,726	64
	3rd "	9,632	69
	4th "	9,505	81
1952	1st "	9,405	98
	2nd "	9,271	101

Calculate the coefficient of correlation and comment on the result.
Institute of Statisticians.

8. Calculate a coefficient of correlation for the following two sets of figures. How would you interpret your result?

	Unemployment in Great Britain: month averages (000)	Company's annual turnover (£000)
1954	285	274
1955	232	297
1956	257	284
1957	312	317
1958	457	335
1959	475	386
1960	360	355

9. Two judges in a contest were asked to rank 8 candidates in their order of preference. The rankings were as follows:

Candidate	Judge A	Judge B
One	5	4
Two	2	5
Three	8	7
Four	1	3
Five	4	2
Six	6	8
Seven	3	1
Eight	7	6

Find the coefficient of rank correlation and say to what extent there is agreement between the two judges.

10. (a) Calculate the coefficient of correlation for the following data:

Firm	Annual percentage increase in advertising expenditure	Annual percentage increase in sales revenue
A	1	1
B	3	2
C	4	4
D	6	4
E	8	5
F	9	7
G	11	8
H	14	9

(b) What is the purpose of finding the correlation coefficient and what does its value indicate in respect of the above data on advertis-ing expenditure and sales revenue?

Association of Certified Accountants.

11. Two expert advisers rank ten products of a firm in order of reliability in use, as in the following table:

	Product:	I	II	III	IV	V	VI	VII	VIII	IX	X
	Products (ranked in order of reliability)										
Advisers:	A	6	2	8	10	7	3	1	4	9	5
	B	5	2	7	9	6	3	1	8	10	4

Calculate the rank correlation coefficient and indicate whether you think that it suggests significant agreement between the views of the two advisers.

Institute of Chartered Secretaries and Administrators.

CHAPTER 18

Regression

THE LINE OF "BEST FIT"

In Chapter 17 the subject of linear correlation was discussed and Fig. 67 gives an example. It was there shown that the points giving the ages of husbands against the corresponding ages of their wives fell either side of a straight line and that this line was described as a line of "best fit" — a kind of average line.

The object of this chapter is to describe how this line can be drawn, to show how its formula can be obtained and to see, further, what is meant by this line of "best fit".

THE EQUATION $y = bx + a$

In Chapter 11, dealing with graphs, it was pointed out that x is the independent variable, that the value of y will depend upon the values given to x and that y is the dependent variable. Provided we know a and b, we can find the value of y for any given value of x. The following equations are examples of equations of the form $y = bx + a$.

$$y = 1.2x - 2$$
$$y = -6x + 4$$
$$y = 0.8x + 0.7.$$

All these equations when shown in the form of graphs will be straight lines. The graph of the first equation would cut the y-axis at a point -2 units below the origin. This is fairly obvious, since when $x = 0$, $y = -2$. Similarly, the graph of the second equation cuts the y-axis at a point 4 units above the origin and in the third example 0.7 units above the origin. In general, the graph of the equation $y = bx + a$ will cut the y-axis a units above the origin (or, if a is a minus quantity, below).

In the first of the equations just given the value of y becomes 1.2 units more with every increase in the value of x by 1. 1.2 is the value of the slope of the line (also known as the tangent of the angle of slope). In the second equation the slope is negative, that is, as shown in Fig. 68, an example of inverse correlation (as x gets bigger y gets smaller). In general, the graph of the equation $y = bx + a$ has a slope whose value is b.

In Fig. 71 (a) the line shown has the equation $y = 1.1x + 0.7$. It can be seen that the line cuts the y-axis at a point 0.7. The slope can be calculated as follows:

when $x = 3, y = 4$
when $x = 5, y = 6.2$

Difference in value of y = 2.2

Difference in value of x = 2

$$\text{Slope} = \frac{2.2}{2} = 1.1.$$

THE REGRESSION EQUATION

This is the equation which gives the relationship between variables when there is not a "unique" relationship between them. Thus, in the case of husbands' and wives' ages, given the age of the husband, it is not possible to say what the age of his wife will be, but it will be possible to estimate by giving the average age of wives whose husbands are of a given age. When the correlation is linear the regression equation will be of the form

$$y = bx + a.$$

The husband's age could be represented by x. It would be necessary to find the value of a and b; then for any given age of the husband the average age of wives of men of that age could be calculated. In any particular case the actual age of the wife might be different from the computed value. The sum of the deviations between actual and computed ages would be "nil". The line is therefore an "average" value line. This is what is meant by being a line of "best fit".

THE REGRESSION OF y ON x

So far we have considered the case of "given x, what is the value of y?"; x is independent. This is also known as the regression of y on x, and b in the equation $y = bx + a$ is the regression coefficient. It shows that y changes b times as fast as x.

THE REGRESSION OF x ON y

It is also possible to make y independent. Given the age of a wife, what is the age of her husband? The regression equation in this case is

$$x = b_1 y + a_1.$$

It is important to note that this line will *not* coincide with the regression equation of $y = bx + a$ unless there is complete correlation between the two variables, and in this case, of course, there would be no deviations from the regression line and the points of the scattergram would all lie on the straight line.

CALCULATIONS TO ESTABLISH REGRESSION EQUATIONS
Example 1.

Number of men employed on project	Total output in units
x	y
1	1
2	3
3	5
4	6
5	5

It is required to find:

(*a*) the regression of y on x;
(*b*) the regression equation of y on x;
(*c*) the regression of total output on number of men employed;
(*d*) the regression line $y = bx + a$.

These are four ways of saying the same thing. It is necessary to find b and a. Then given any number of men employed, the output can be estimated. It is also proposed to find the equation of $x = b_1 y + a_1$ so that, given any output, we can estimate the number of men required on the project.

The calculations required are as follows:

x	y	x^2	y^2	xy
1	1	1	1	1
2	3	4	9	6
3	5	9	25	15
4	6	16	36	24
5	5	25	25	25
15	20	55	96	71

$\Sigma x = 15; \Sigma y = 20; \Sigma x^2 = 55;$
$\Sigma y^2 = 96; \Sigma xy = 71.$
$N = 5$ (there are 5 observations).

These figures are precisely the same figures that must be computed to find the coefficient of correlation. If only the regression of y or x is required it is not necessary to calculate Σy^2.

THE REGRESSION EQUATION OF Y ON X
The equation required is of the form

$$y = bx + a.$$

It is required to find the values of b and a. This necessitates solving the following simultaneous equations:

(a) $\Sigma y = b\Sigma x + na$
(b) $\Sigma xy = b\Sigma x^2 + a\Sigma x.$

In the example we are dealing with these are:

$$20 = 15b + 5a \tag{1}$$
$$71 = 55b + 15a \tag{2}$$

Multiplying (1) by 3,

$$60 = 45b + 15a \tag{3}$$

Subtracting (3) from (2),

$$11 = 10b$$
$$\therefore \quad b = 1.1.$$

Substitute $b = 1.1$ in equation (1)

$$20 = 16.5 + 5a$$

$$\therefore \quad a = \frac{3.5}{5} = 0.7.$$

The required equation is $y = 1.1x + 0.7$ (see Fig. 71(a)).

THE REGRESSION EQUATION OF X ON Y
The equation required is of the form

$$x = b_1 y + a_1.$$

It is required to find the values of b_1 and a_1. This necessitates solving the following simultaneous equations:

(a) $\Sigma x = b_1\Sigma y + na_1$
(b) $\Sigma xy = b_1\Sigma y^2 + a_1\Sigma y.$

In the example we are dealing with these are:

$$15 = 20b_1 + 5a_1 \tag{1}$$
$$71 = 96b_1 + 20a_1 \tag{2}$$

Multiplying (1) by 4,

$$60 = 80b_1 + 20a_1 \tag{3}$$

Deducting (3) from (2),

$$11 = 16b_1$$

$$\therefore \quad b_1 = \frac{11}{16} = 0.6875.$$

Substituting for b_1 in (1),

$$a_1 = \frac{15 - (20 \times 0.6875)}{5} = 0.25.$$

The required regression equation is

$$x = 0.6875y + 0.25 \ (see \ \text{Fig. 71 } (b)).$$

PROCEDURE

Establishing linear regression equations

1. Compute X^2, Y^2, XY.
2. Compute ΣX, ΣY, ΣX^2, ΣY^2, ΣXY. (The first two steps are the same as those in the computation of the coefficient of correlation).
3. (a) For the regression equation of Y on X, i.e. $y = bx + a$, solve the following simultaneous equations:
    (i) $\Sigma Y \ = b\Sigma X + na$
    (ii) $\Sigma XY = b\Sigma X^2 + a\Sigma X$

(b) For the regression equation of X on Y, i.e. $x = b_1 y + a_1$, solve the following simultaneous equations:
    (i) $\Sigma X \ = b_1 \Sigma Y + na_1$
    (ii) $\Sigma XY = b_1 \Sigma Y^2 + a_1 \Sigma Y$.

4. Substitute the values of a, a_1, b, b_1 in the regression equations.

CHANGING THE REGRESSION EQUATION

Compare the variables y and x in the first two columns in Example 2 below with the variables u and v in the next two columns.

The variable u is obtained by subtracting 64 from the variable y and v is obtained by dividing x by 2. In Chapter 17 on correlation it was shown that the coefficient of correlation between y and x and between u and v were the same. The regression of y on x, however, is *not* the same as the regression of u on v. The origin and the rate of change have altered.

Example 2.

y	x	$u(y - 64)$	$v\left(\dfrac{x}{2}\right)$	uv	v^2
68	18	4	9	36	81
64	16	0	8	0	64
67	20	3	10	30	100
69	24	5	12	60	144
68	22	4	11	44	121
		16	50	170	510

It is proposed to find the regression of u on v and from this regression equation find the regression of y on x.

$$\Sigma u = b\Sigma v + na \qquad (1)$$
$$\Sigma uv = b\Sigma v^2 + a\Sigma v \qquad (2)$$
$$16 = 50b + 5a \qquad (1)$$
$$170 = 510b + 50a \qquad (2)$$

$$b = 1 \text{ and } a = -\frac{34}{5}.$$

The regression equation of u on v is $u = 1v - \dfrac{34}{5}$.

Substituting for u, $y - 64$ and for v, $\dfrac{x}{2}$ this equation becomes

$$y - 64 = \tfrac{1}{2}x - \frac{34}{5}$$

$$\therefore \ y = \tfrac{1}{2}x + 57\tfrac{1}{5}.$$

This is the regression equation of y on x.

THE REGRESSION COEFFICIENT

The slope of the regression line is known as the regression coefficient. It is the value of b and b_1 in the regression equations.

REGRESSION AND CORRELATION

Figures 71 (a) and 71 (b) show the graphs of the two regression lines. If they were superimposed it would be seen that the angle between them is very small, denoting a great deal of correlation between the two variables.

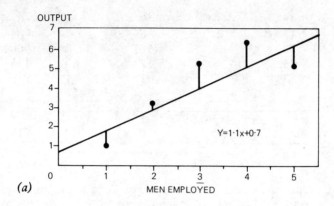

(a)

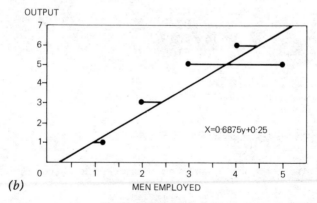

(b)

Fig. 71. (a) Regression of y on x.
(b) Regression of x on y.

The coefficient of correlation is in fact the geometric mean of the two regression coefficients, and is therefore $\sqrt{1.1 \times 0.6875} = 0.87$.

The following calculations will show that the deviations from the line of "best fit" add up to nil. Actual measurements on the graph will show the same thing.

Example 3.

$$y = 1.1 x + 0.7$$

x	y (actual)	y (computed from equation)	Deviation
1	1	1.8	−0.8
2	3	2.9	+0.1
3	5	4.0	+1.0
4	6	5.1	+0.9
5	5	6.2	−1.2

$$x = 0.6875y + 0.25$$

y	x (actual)	x (computed from equation)	Deviation
1	1	0.9375	+0.0625
3	2	2.3125	−0.3125
5	3	3.6875	−0.6875
6	4	4.3750	−0.3750
5	5	3.6875	+1.3125

ALTERNATIVE METHOD OF CALCULATING THE REGRESSION EQUATION

$$y = bx + a$$

$$b = \frac{\Sigma xy - \dfrac{(\Sigma x)(\Sigma y)}{n}}{\Sigma x^2 - \dfrac{(\Sigma x)^2}{n}}$$

$$= \frac{71 - \dfrac{(15)(20)}{5}}{55 - \dfrac{15^2}{5}}$$

$$= \frac{11}{10} = 1.1 \text{ (as before)}.$$

$$a = \frac{(\Sigma x)(\Sigma xy) - (\Sigma y)(\Sigma x^2)}{(\Sigma x)^2 - n(\Sigma x^2)}$$

$$= \frac{(15)(71) - (20)(55)}{(15)(15) - (5)(55)}$$

$$= \frac{-35}{-50} = 0.7 \text{ (as before).}$$

LIMITS IN THE INTERPRETATION OF A REGRESSION LINE

Interpolation between the extreme values of the observations given is normally quite safe. However, *extrapolation outside* the extreme values is often quite meaningless or dangerous. As an example of this, according to our equation, when no men are employed output is 0.7 units.

QUESTIONS

1. Plot the values in the following table on squared paper, and join them with a smooth curve:

x	y
1	10
2	19
3	27
5	40
6	45
7	49
8	52
9	54
11	55

(a) From your graph estimate: (i) the value of y when $x = 4$; (ii) the value of y when $x = 10$; (iii) the maximum value of y, and the value of x which corresponds to this value of y.

(b) Calculate the value of a and b and draw on your graph the "line of best fit" $y = a + bx$. Using this line, what values of y do you get for $x = 4$ and $x = 10$?

(*Note:* The maximum value of y on the smooth curve is greater than 55. It reaches its highest value between 9 and 11 and has already started to fall before it becomes equal in value to 55 at the point when $x = 11$.)

Institute of Cost and Management Accountants.

2. Calculate the values of m and k for the equation $y = mx + k$ to show the regression of profit per unit of output on output.

Output (thousands)	Profit per unit of output (£)
5	1.7
7	2.4
9	2.8
11	3.4
13	3.7
15	4.4

Estimate from your equation the profit per unit of output when there is an output of 10,500.

3. The following figures refer to 10 specimens of a light metal alloy. The first row of figures gives the measure of hardness and the second row the corresponding tensile strength. You are required to find an equation from which an estimate of the tensile strength can be determined from a known hardness (assume linear relationship).

Hardness	22.9	17.8	20.8	21.3	20.7
Tensile strength	4.4	4.6	5.6	4.2	3.2
Hardness	20.9	17.5	13.6	23.3	18.1
Tensile strength	4.7	3.8	4.5	4.2	5.2

Institute of Statisticians.

4. From the data below fit a regression line of the form $y = ax + b$.

Year	Sales (£000)
1975	1
1976	3
1977	5
1978	6
1979	5

On the assumption that sales will continue to grow in the same way, estimate sales for 1981 (to nearest £ thousand).

5. The following table shows the amount of electricity generated by steam plant in Barbalonia in the ten years, 1970 to 1979, and the amount of coal used in generation:

Year	Electricity generated (x) (thousand million kilowatt hours)	Coal used (y) (million tonnes)
	(x)	(y)
1970	37	23
1971	41	26
1972	43	27
1973	47	29
1974	49	30
1975	55	33
1976	60	35
1977	62	36
1978	65	37
1979	72	40

Find the equations of regression for y on x and x on y.

6. From the figures below obtain a formula giving pig-iron production in terms of coal consumption.

Year	Coal consumed in blast furnaces (in units of 100,000 tonnes)	Pig-iron produced (in units of 100,000 tonnes)
1973	14.51	7.59
1974	11.69	6.19
1975	7.11	3.77
1976	6.53	3.57
1977	7.37	4.14
1978	10.47	5.97
1979	10.79	6.42

7. From the information given below calculate the linear regression equation of annual tonne-kilometres per vehicle (thousands) *on* average vehicle carrying capacity (tonnes).

Let Y be the annual tonne-kilometres per vehicle (thousands) and X the average vehicle carrying capacity (tonnes).

Comment on your calaculated regression equation.

Year	Annual tonne-km per vehicle (thousands) (Y)	Average carrying capacity (tonnes) (X)
1969	80.1	4.7
1970	83.7	4.9
1971	87.1	5.1
1972	87.5	5.3
1973	98.3	5.5
1974	103.7	5.8
1975	111.9	6.1
1976	111.1	6.3
1977	119.9	6.6
1978	127.4	6.9
1979	129.3	7.2

Given that $\Sigma Y = 1{,}140$ $\Sigma X = 64.4$

$\Sigma X^2 = 384$ $\Sigma XY = 6{,}820.55$

CHAPTER 19

Probability and Probability Distributions

STATISTICAL PROBABILITY

THE PROBABILITY OF AN EVENT OCCURRING

If an action can have any one of n equally likely outcomes, and of these r would produce an event E, the probability of E is defined as $\dfrac{r}{n}$.

Example 1. A fair six-sided die is given a fair throw. There are six possible equally likely ways it can fall. Of these six possible outcomes three would result in an odd number. The probability of obtaining an odd number is therefore $\frac{3}{6} = \frac{1}{2}$:

Example 2. From a bowl containing 100 red balls and 50 white balls, a ball is drawn at random. What is the probability it will be red?

If the ball is drawn at random (remember the technical sense of the word random) the action can have 150 equally likely results—any one of the 150 balls can be chosen. Of these results 100 would produce a red ball. The probability of drawing a red ball is therefore $\frac{100}{150} = \frac{2}{3} = 67\%$.

In betting language the odds of getting a red ball are two to one.

THE PROBABILITY OF AN EVENT NOT OCCURRING

This is one minus the probability of its occurrence. In symbols this is $P(\overline{E}) = 1 - P(E)$.

Example 3. What is the probability of not getting six in a fair throw of a fair six-sided die?

The probability of getting a six is $\frac{1}{6}$. The probability of not getting a six will be $1 - \frac{1}{6} = \frac{5}{6}$. The answer to the question would be set out as follows:

$$P(6) = \frac{1}{6}$$

$$P(\overline{6}) = 1 - \frac{1}{6} = \frac{5}{6}$$

THE MEASUREMENT OF PROBABILITY

From the definition of probability it can be seen that it can vary between 1 (certainty) and 0 (certainty that the event will *not* happen). Probabilities are often expressed as a percentage. In Example 1 the probability of getting an odd number is 50 per cent; in Example 3, the probability of not getting a six is 83 per cent.

An event can either occur or not occur. There is no other possibility. Hence it is *certain* that an event either occurs or does not occur. The probability of an event $P(E)$ and the probability of its non-occurrence $P(\overline{E})$ must therefore add up to one.

CALCULATING PROBABILITY

THE ADDITION RULE OF PROBABILITY

If there are a number of events and they are *mutually exclusive* (that is, if any one happens the other cannot), then the probability of *any one* of them happening is the sum of their separate probabilities. In symbols this will be:

$$P(E_1 \ or \ E_2 \ or \ E_3 \ldots E_n) = P(E_1) + P(E_2) + P(E_3) + \ldots P(E_n).$$

Example 4. A six-sided die is thrown into the air. What is the probability that *either* a three *or* a four will fall uppermost?

Both these events are *mutually exclusive*. (If a three turns up, a four cannot; if a four turns up, a three cannot.) The probability of *either* a three *or* a four turning up is therefore the sum of their separate probabilities:

$$P(3) = \frac{1}{6} \qquad\qquad P(4) = \frac{1}{6}$$

$$P(3 \ or \ 4) = \frac{1}{6} + \frac{1}{6} = \frac{1}{3} \ .$$

THE MULTIPLICATION RULE OF PROBABILITY

If there are a number of events and they are *independent* (that is, the occurrence of one event cannot affect the occurrence of the others), then the probability of their *all* occuring is the product of their separate probabilities. In symbols this will be:

$$P(E_1 \text{ and } P_2 \text{ and } P_3 \text{ and} \ldots P_n) =$$
$$P(E_1) \times P(E_2) \times P(E_2) \times \ldots P(E_n).$$

Example 5. A coin is tossed into the air and a card is drawn from a well shuffled pack of cards. What is the probability that *both* a head falls uppermost and an ace is drawn?

Both these events are *independent* (the throwing of a coin cannot affect the drawing of a card). The probability of *both* events occurring is the product of their separate probabilities:

$$P(\text{Head}) = \tfrac{1}{2} \qquad\qquad P(\text{Ace}) = \tfrac{4}{52} = \tfrac{1}{13}$$
$$P(\text{Head } and \text{ Ace}) = \left(\tfrac{1}{2}\right)\left(\tfrac{1}{13}\right) = \tfrac{1}{26}.$$

THE MULTIPLICATION RULE APPLIED TO CONDITIONAL PROBABILITY

When events are *dependent*, then the probability of any one event occurring depends on whether the others have occurred.

Example 6. From a bowl containing 100 red balls and 50 white balls two balls are drawn at random. What is the probability they will both be red?

The probability of the first ball being red is $\tfrac{100}{150}$ *(see* Example 2*)*. When the first ball is drawn and *if it is red*, there will be 99 red balls and 50 white balls left in the bowl. The probability of the second ball being red is therefore $\tfrac{99}{149}$. The probability of the first ball being red *and* the second ball being red is therefore

$$\left(\frac{100}{150}\right)\left(\frac{99}{149}\right) = \frac{198}{447} = 44.3\%.$$

COMBINATIONS

Consider four objects, say buttons, and in order to identify them let them be referred to as A, B, C and D. The following combinations (selections) of two buttons can be chosen: AB, AC, AD, BC, BD and CD (6 selections). No regard is paid to the order, AB being the same combination as BA.

Note. If order were taken into account it would be a question of *permutations* not combinations. Often in everyday speech the word permutation is incorrectly used when combination is meant.

Example 7. In how many ways can two buttons be chosen from four?

The number of ways is $\dfrac{(4)\,(3)}{(1)\,(2)} = 6$. $\dbinom{4}{2}$ means the number of combinations of two things obtainable from four: $\dbinom{4}{2} = 6$.

Note. Do not, as many students do, put a division line between the 4 and the 2.

Example 8. How many different selections of 8 matches is it possible to get from 10 matches?

$$\binom{10}{8} = \frac{(10)(9)(8)(7)(6)(5)(4)(3)}{(1)(2)(3)(4)(5)(6)(7)(8)} = 45.$$

Another way of selecting 8 matches from 10 would be to chose 2 and reject them, leaving a selection of 8. The number of ways of selecting 2 matches from 10 will therefore be the same as the number of ways of selecting 8:

$$\binom{10}{8} = \binom{10}{2}$$

$$\binom{10}{2} = \frac{(10)(9)}{(1)(2)} = 45.$$

The number of combinations of r things from n unlike things is denoted by:

$\dbinom{n}{r}$ which equals $\dfrac{(n)(n-1)(n-2)(n-3)\ldots(n-r+1)}{(1)(2)(3)(4)\ldots(r)}$.

It is also equal to the number of combinations of $n-r$ things taken from n unlike things denoted by $\dbinom{n}{n-r}$. Example 8 exemplifies this ($n = 10, r = 8, n - r = 2$).

Example 9. Samples of four are taken from a continuing supply of buttons, some of which are plain and some are fancy. In how many ways can a sample have no plain buttons, 1, 2, 3 and 4 plain button?

There is only one way in which the sample of four can have no plain buttons and that is when they are all fancy. $\dbinom{4}{0} = 1$.

We compute the number of combinations as follows.

(a) Number of ways the sample can have no plain buttons:

$$\binom{4}{0} = 1.$$

(b) Number of ways the sample can have 1 plain button:

$$\binom{4}{1} = 4.$$

(c) Number of ways the sample can have 2 plain buttons:

$$\binom{4}{2} = \frac{(4)(3)}{(1)(2)} = 6.$$

(d) Number of ways the sample can have 3 plain buttons:

$$\binom{4}{3} = \frac{(4)(3)(2)}{(1)(2)(3)} = 4.$$

A better answer would be $\binom{4}{3} = \binom{4}{4-3} = \binom{4}{1} = 4.$

(e) Number of ways the sample can have 4 plain buttons:

$$\binom{4}{4} = 1 \quad (\text{i.e. } \binom{n}{n} = 1).$$

The answer to Example 9 is shown in diagrammatic form in Fig. 72.

PROBABILITY OF OBTAINING r ITEMS WITH A GIVEN CHARACTERISTIC IN A SAMPLE OF n

Consider the case of a sample of 4 taken from a continuing supply of plain and fancy buttons; it is required to calculate the probability of the sample having two plain buttons.

It is first necessary to know if just one button were selected instead of 4, the probability of its being plain. Further information is therefore needed concerning the population from which the samples are chosen. Suppose the manufacturer produces $\frac{4}{5}$ths of his output plain buttons and $\frac{1}{5}$th fancy buttons; the probability of a single button being plain would be $\frac{4}{5}$.

The probability of a button being plain is $\frac{4}{5}$.

p (the probability that a single item possesses the characteristic).

The probability of a button *not* being plain is $\frac{1}{5}$.

q (the probability that a single item *does not* possess the characteristic).

$$q = 1 - p$$

$P(\text{1st button plain}) = \frac{4}{5}$. $\qquad p$

$P(\text{1st and 2nd buttons plain}) =$

$$\left(\frac{4}{5}\right)\left(\frac{4}{5}\right) = \left(\frac{4}{5}\right)^2 . \qquad p^r$$

Number of combination	r	Selections	n = 4				Number of combinations
			A	B	C	D	
1	0		♠	♠	♠	♠	$\binom{4}{0} = 1$
1		A	⊙	♠	♠	♠	
2	1	B	♠	⊙	♠	♠	$\binom{4}{1} = 4$
3		C	♠	♠	⊙	♠	
4		D	♠	♠	♠	⊙	
1		AB	⊙	⊙	♠	♠	
2		AD	⊙	♠	♠	⊙	
3	2	CD	♠	♠	⊙	⊙	$\binom{4}{2} = 6$
4		BD	♠	⊙	♠	⊙	
5		AC	⊙	♠	⊙	♠	
6		BC	♠	⊙	⊙	♠	
1		BCD	♠	⊙	⊙	⊙	
2	3	ACD	⊙	♠	⊙	⊙	$\binom{4}{3} = 4$
3		ABD	⊙	⊙	♠	⊙	
4		ABC	⊙	⊙	⊙	♠	
1	4	ABCD	⊙	⊙	⊙	⊙	$\binom{4}{4} = 1$

⊙ PLAIN ♠ FANCY

Fig. 72. Combinations: n = 4; r = 0, 1, 2, 3 and 4.

P(1st *and* 2nd buttons plain) =

$$\left(\frac{4}{5}\right)\left(\frac{4}{5}\right) = \left(\frac{4}{5}\right)^2 . \qquad\qquad p^r$$

P(1st *and* 2nd buttons plain *and*
3rd *and* 4th buttons *not* plain) =

$$\left(\frac{4}{5}\right)\left(\frac{4}{5}\right)\left(\frac{1}{5}\right)\left(\frac{1}{5}\right) = \left(\frac{4}{5}\right)^2\left(\frac{1}{5}\right)^2 . \qquad q^{(n-r)} p^r$$

There are six ways of having two
plain buttons in a sample of four
(*see* Fig. 72).

P(2 plain buttons in a sample of 4) =

$$\binom{4}{2}\left(\frac{4}{5}\right)^2\left(\frac{1}{5}\right)^2 = \frac{96}{625}. \qquad \binom{n}{r} q^{(n-r)} p^r$$

Example 10. The output of a button manufacturer is $\frac{4}{5}$ths plain button and $\frac{1}{5}$th fancy. If samples of 4 are taken at random what is the probability of obtaining (*a*) 0, (*b*) 1, (*c*) 2, (*d*) 3, (*e*) 4 plain buttons in a sample?

(*a*) $P(0) = \binom{4}{0}\left(\frac{1}{5}\right)^4 = \dfrac{1}{625}$.

(*b*) $P(1) = \binom{4}{1}\left(\frac{1}{5}\right)^3\left(\frac{4}{5}\right) = \dfrac{16}{625}$.

(*c*) $P(2) = \binom{4}{2}\left(\frac{1}{5}\right)^2\left(\frac{4}{5}\right)^2 = \dfrac{96}{625}$.

(*d*) $P(3) = \binom{4}{3}\left(\frac{1}{5}\right)\left(\frac{4}{5}\right)^3 = \dfrac{256}{625}$.

(*e*) $P(4) = \binom{4}{4}\left(\frac{4}{5}\right)^4 = \dfrac{256}{625}$.

From this information many more questions can easily be answered:

$$P(1 \textit{ or } 2 \text{ plain buttons}) = \frac{16}{625} + \frac{96}{625} = \frac{112}{625}$$

P(less than 4 plain buttons) = P(0 *or* 1 *or* 2 *or* 3)

$$= 1 - P(4) = 1 - \frac{256}{625} = \frac{369}{625}$$

$$P(0 \text{ or } 1 \text{ or } 2 \text{ or } 3 \text{ or } 4) = \frac{1 + 16 + 96 + 256 + 256}{625} = \frac{625}{625} = 1$$

Since it is certain that the sample will contain either 0 *or* 1 *or* 2 *or* 3 *or* 4 plain buttons (there is no other possibility) the separate probabilities *must* add up to 1 indicating certainty.

THE BINOMIAL DISTRIBUTION

Referring to Example 10, the probabilities of 0, 1, 2, 3 and 4 plain buttons being found in the sample are:

$$\left(\frac{1}{5}\right)^4; \quad \binom{4}{1}\left(\frac{1}{5}\right)^3\left(\frac{4}{5}\right); \quad \binom{4}{2}\left(\frac{1}{5}\right)^2\left(\frac{4}{5}\right)^2; \quad \binom{4}{3}\left(\frac{1}{5}\right)\left(\frac{4}{5}\right)^3; \quad \left(\frac{4}{5}\right)^4.$$

These probabilities are the same as the terms in the expansion of the binomial $\left(\frac{1}{5} + \frac{4}{5}\right)^4$. Note that $\frac{1}{5}$th is the probability of *not* getting a plain button (denoted by q) and $\frac{4}{5}$ths is the probability of getting a plain button—plain being the characteristic in question (denoted by p)—and that $\frac{1}{5} + \frac{4}{5} = 1$ (i.e. $q + p = 1$); 4 is the number in the sample (denoted by n).

The binominal distribution can be expressed by the following equation:

$$(q + p)^n = q^n + \binom{n}{1}q^{(n-1)}p^1 + \binom{n}{2}q^{(n-2)}p^2 + \ldots p^n$$

where:

q^n is the probability of there being 0 items having the given characteristic in a sample size of n;

$\binom{n}{1}q^{(n-1)}p$ is the probability of there being 1 such item;

$\binom{n}{2}q^{(n-2)}p^2$ is the probability of there being 2 such items and

so on, and where:

p^n is the probability that all n items in the sample will have the given characteristic.

The *distribution of the probabilities* of obtaining 0, 1, 2, 3 ... n items with a given characteristic in random samples of n items (in Example 10, 0, 1, 2, 3 and 4 plain buttons in samples of 4) is a *binomial* distribution *provided* that the probability of obtaining a single item with the given characteristic remains constant each time an item is drawn from the population. This will be so only if the item is replaced after each drawing (*sampling with replacement*) or the population is infinite or can be considered so.

Where the population is being continually maintained and the items with the given characteristic are a constant proportion of the supply (e.g. the continuing supply of Example 9) the distribution of the probabilities may be considered a binomial distribution. It is for all practical purposes sampling with replacement.

Again where the population is large, sampling without replacement is practically identical to sampling with replacement and the population can be treated as if it were infinite. Suppose that instead of the continuing supply of the previous example there were only 100 buttons—80 plain and 20 fancy—then for samples of 4 *without replacement* the probabilities of 0, 1, 2, 3 and 4 plain buttons would be as follows.

$$P(0) = 1 \left(\frac{20}{100}\right)\left(\frac{19}{99}\right)\left(\frac{18}{98}\right)\left(\frac{17}{97}\right) = 0.0012 \text{ (as against 0.0016)}$$

$$P(1) = 4 \left(\frac{80}{100}\right)\left(\frac{20}{99}\right)\left(\frac{19}{98}\right)\left(\frac{18}{97}\right) = 0.0233 \text{ (as against 0.0256)}$$

$$P(2) = 6 \left(\frac{80}{100}\right)\left(\frac{79}{99}\right)\left(\frac{20}{98}\right)\left(\frac{19}{97}\right) = 0.1531 \text{ (as against 0.1536)}$$

$$P(3) = 4 \left(\frac{80}{100}\right)\left(\frac{79}{99}\right)\left(\frac{78}{98}\right)\left(\frac{20}{97}\right) = 0.4191 \text{ (as against 0.4096)}$$

$$P(4) = 1 \left(\frac{80}{100}\right)\left(\frac{79}{99}\right)\left(\frac{78}{98}\right)\left(\frac{77}{97}\right) = 0.4033 \text{ (as against 0.4096)}$$

$$\overline{1.0000} \qquad \overline{1.0000}$$

The difference between sampling without replacement and sampling with replacement with a finite population of only 100 was very small and the population soon becomes large enough for the difference to become negligible.

THE MEAN AND STANDARD DEVIATION OF THE BINOMIAL DISTRIBUTION

Referring once more to the button example it is seen that the *number* of plain buttons and the *proportion* of plain buttons varies from sample to sample. The distributions of the numbers and proportions can be described by computing their means and standard deviations.

Example 11. The output of a button manufacturer is $\frac{4}{5}$ths plain buttons and $\frac{1}{5}$th fancy. Samples of 4 are taken. What is *(a)* the average number of plain buttons in a sample and the standard deviation, and *(b)* the average proportion of plain buttons and the standard deviation of the proportion of plain buttons in a sample.

(a)

Number of plain buttons x	Number of samples f	fx	fx^2
0	1	0	0
1	16	16	16
2	96	192	384
3	256	768	2,304
4	256	1,024	4,096
	625	2,000	6,800

$$\bar{x} = \frac{\Sigma fx}{\Sigma f} = \frac{2,000}{625} = 3.2$$

$$\text{S.D.} = \sqrt{\frac{\Sigma fx^2}{n} - \bar{x}^2}$$

$$= \sqrt{\frac{6,800}{625} - 3.2^2} = 0.8$$

(b)

Proportion of plain buttons x	Number of samples f	fx	fx^2
$0/4$	1	0	0
$1/4$	16	4	1
$1/2$	96	48	24
$3/4$	256	192	144
$4/4$	256	256	256
	625	500	425

$$\bar{x} = \frac{500}{625} = \frac{4}{5} = 0.8$$

$$\text{S.D.} = \sqrt{\frac{425}{625} - 0.8^2} = 0.2$$

The probabilities of obtaining 0, 1, 2, 3 and 4 plain buttons have already been computed in Example 10. If there is a probability of $\frac{1}{625}$ th of obtaining no plain buttons in a sample then in 625 samples there is a probability of there being 1 sample of the 625 which has no plain buttons; if there is a probability of $\frac{16}{625}$ ths of obtaining 1 plain button in the sample there is a probability that 16 of the 625 samples contain one plain button and so on.

It is seen that by dividing the frequency of each value of the variable by the sum of the frequencies, the probability of each value of the variable occurring is obtained.

$$P(x_i) = \frac{f_i}{\Sigma f} \qquad \text{(where } x_i \text{ is a given value of the variable and } f_i \text{ is its frequency).}$$

The mean and standard deviation of a binomial distribution can also be calculated from formulae.

(a) The *mean number* of items possessing a given characteristic in a sample of size n where the proportion of such items in the population is p is pn.

(b) The *mean proportion* of items in the sample is p.

(c) The *standard deviation of the number* is $\sqrt{pqn}$.

(d) The *standard deviation of the proportion* is $\sqrt{\dfrac{pq}{n}}$

Using these formulae for Example 11 the calculations are as follows.

(a) Mean number of plain buttons $(pn) = \frac{4}{5} \times 4 = 3.2$

(b) Mean proportion of plain buttons $(p) = \frac{4}{5} = 0.8$

(c) S.D. of number of plain buttons $(\sqrt{pqn}) = \sqrt{\left(\frac{4}{5}\right)\left(\frac{1}{5}\right)(4)} = 0.8$

(d) S.D. of proportion of plain buttons $\left(\sqrt{\dfrac{pq}{n}}\right) = \sqrt{\dfrac{\left(\frac{4}{5}\right)\left(\frac{1}{5}\right)}{4}} = 0.2$

It will be realised that there will be a whole series of binomial distributions depending on the values of p and n. These are the *parameters* of a binomial distribution.

THE NORMAL DISTRIBUTION

Unlike the binomial distribution which is concerned with discrete values of the variable (e.g. 0, 1, 2 plain buttons) the "normal" distribution is an example of a continuous distribution. Continuous variables can have any value (e.g. height, time, pressure, volume, weight). Given any two measurements, however close, it is always possible to find a third which lies between them. *A continuous variable has an infinite number of values.*

Much data are "normally" distributed. Intelligence quotients, heights and errors made in observations are examples. As far as business is concerned its great importance lies in large sampling theory (dealt with in Chapter 20) and its use as a very close approximation to the binomial distribution (dealt with later in this chapter).

Note. The word "normal" as used in normal distribution does not mean usual or typical. The normal distribution is one particular distribution and would best be called by its other but very much less used name, the Gaussian distribution.

The frequency curve of a "normal" distribution is shown in Fig. 73. It is a symmetrical, bell-shaped figure and slopes downwards

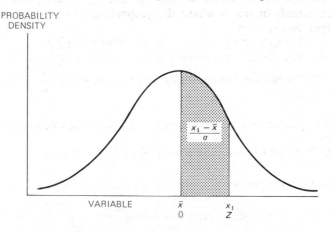

Fig. 73. *The "normal" distribution and standardised deviates.*

either side of the mean and approaches the x-axis in both directions without touching it. (For most practical purposes it can be considered to meet the axis four standard deviations either side of the mean.) The ordinate at $\bar{x}$ bisects the area under the curve.

If the frequencies are transformed into probabilities the total area under the curve becomes unity and the y-axis shows the probability density. In the same way as the frequencies of discrete values are areas of a histogram, so probabilities of continuous values are areas. However, in the case of a continuous distribution it is only possible to calculate probabilities over a range of values; the range can be, however, as small as is required.

To find the probability of any range of values which are normally distributed it is necessary to know the mean value of the distribution (denoted by $\bar{x}$) and the standard deviation (denoted by σ). These are the *parameters* of the "normal" distribution. The values are converted to *standardised deviates* (denoted by Z).

$$Z = \frac{x - \bar{x}}{\sigma}$$

This means that the mean of the normal distribution becomes 0 and the deviations from the mean are measured in units of standard deviations.

The probabilities of values between the mean and Z are shown in the table in Appendix C headed *Area under the normal curve*. This is the shaded area in Fig. 73.

USE OF THE TABLE—AREA UNDER NORMAL CURVE (Appendix C).

Remember the total area under the curve is unity and that the ordinate at 0 (the mean) bisects the area.

Note. It is immaterial whether Z has a minus or positive value; the area is exactly the same.

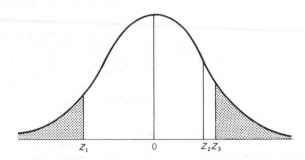

Fig. 74. Areas under the "normal" curve.

Probability of values between Z_2 and Z_3 = area for Z_3 *less* area for Z_2.

Probability of values exceeding Z_3 = 0.5 *less* area for Z_3.
(This area—1 tail—is shaded on diagram.)

Probability of values not exceeding Z_3 = 0.5 *plus* area for Z_3.

Probability of values between Z_1 and Z_3 = area for Z_1 *plus* area for Z_3.

Probability of values less than Z_1 = 0.5 *less* area for Z_1.
(This area—a tail-is shaded on diagram.)

Probability of a value less than Z_1 *or* = 1 *less* area for Z_1 greater than Z_3 *less* area for Z_3.

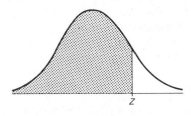

Fig. 75. Different tables show different areas under the "normal" curve.

Note. Not all tables of areas under the "normal" curve are as shown in Appendix C. Some give the shaded area of Fig. 75. Areas in such tables would be 0.5 greater than the figures given in this book. Others show a tail—the blank part of Fig. 75. Areas in those tables would be 0.5 *less* the areas given in this book.

THE "NORMAL" DISTRIBUTION AS AN APPROXIMATION OF THE BINOMIAL

It is often more convenient to treat the binomial distribution as if it were a "normal" distribution provided the following conditions are fulfilled:

(a) n must be large (in practice 30 or more);

(b) neither p nor q must be too close to zero (in practice *both* pn and qn are greater than 5).

Example 12. Twenty per cent of a manufacturer's output is not of the required quality. What is the probability that in a random sample of 30 there will be 9 defective products?

$n = 30$; $pn = 6$ (20% of 30); $qn = 24$ (80% of 30).

The "normal" distribution will be appropriate.

$$\bar{x}(pn) = 6; \sigma (\sqrt{pqn}) = \sqrt{(0.20)(0.80(30)} = 2.19.$$

If the probabilities of obtaining 0, 1, 2 . . . 29 and 30 defectives were shown in the form of a histogram, then the probability of 9 defectives would be shown as a rectangle with the base between 8½ and 9½, the height being the probability density so that the area of the rectangle was its probability. In the same way when using the "normal" distribution approximation, the base of the area represent-the probability of 9 defectives will also be 8½ to 9½ (*see* Fig. 76).

The probability required is the shaded area in Fig. 76.

$$\text{Standardised deviate for 9½} = \frac{9.5 - 6.0}{2.19} = 1.60.$$

$$\text{Standardised deviate for 8½} = \frac{8.5 - 6.0}{2.19} = 1.14.$$

$$P(9 \text{ defectives}) = 0.4452 - 0.3729 = 0.0723.$$

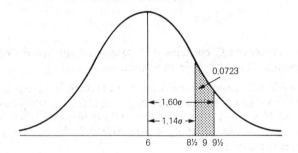

Fig. 76. The "normal" distribution as an approximation of a binomial.

Applying the binomial distribution the probability of 9 defectives would be $\binom{30}{9} (0.80)^{21} (0.20)^9 = 0.0676$ as compared with the "normal" distribution approximation of 0.0723.

THE POISSON DISTRIBUTION

If in a binomial distribution n is large and the probability of the occurrence of an event approaches zero ($p \to 0$) and therefore the probability of its non-occurrence approaches unity ($q \to 1$), the event is known as a *rare event*.

In practice an event is considered rare if n is 50 or more *and* pn is less than 5. The Poisson distribution is then a close approximation to the binomial.

The Poisson distribution is a discrete distribution; the variable can only be an integer, i.e. a whole number. Depicted as a diagram, it would take the form of a histogram. It has only one parameter—the mean (denoted by $m = pn$). Given the mean the distribution is determined.

In a Poisson distribution

$$P(x) = \frac{m^x e^{-m}}{x!}$$

e is a constant: 2.71828

$$e^{-m} = \frac{1}{2.71828^m}$$

$x! = 1 \times 2 \times 3 \times \ldots (x-1)(x)$

$0! = 1$ (read x factorial, 0 factorial . . .)

$x^0 = 1$

Note. Appendix G contains e^{-m} tabulated for certain values of m. Appendix H gives certain factorials.

Example 13. Two per cent of a manufacturer's output has to be rejected for failing to reach the required standard of quality. In a sample of 50 what is the probability of there being 8 or more rejects? Since $n = 50$ and $pn = 0.02 \times 50 = 1.00 < 5$, the Poisson approximation is appropriate: $m = 1$.

Since $m = 1$; $m^x = 1^x = 1$; $e^{-m} = e^{-1} = 0.36788$.

$$P(0) = e^{-1} = 0.36788$$

$$P(1) = \frac{e^{-1}}{1!} = 0.36788$$

$$P(2) = \frac{e^{-1}}{2!} = 0.18394$$

$$P(3) = \frac{e^{-1}}{3!} = 0.06131$$

$$P(4) = \frac{e^{-1}}{4!} = 0.01533$$

$$P(5) = \frac{e^{-1}}{5!} = 0.00306$$

$$P(6) = \frac{e^{-1}}{6!} = 0.00051$$

$$P(7) = \frac{e^{-1}}{7!} = 0.00007$$

$$\overline{0.99998}$$

Probability of 8 or more rejects is $1 - 0.99998 = 0.00002$.

Note that $P(x) = \dfrac{P(x-1)(m)}{x}$.

THE POISSON DISTRIBUTION IN ITS OWN RIGHT

In a given number of items it is possible to count the number that are defective and the number that are not defective. It is then possible to calculate the probability of obtaining a defective item if a single item is drawn at random.

In a given length of cable it is possible to count the number of defects but not the number of no defects. It is possible to count the number of customers entering a shop in a given time but not the number of customers not entering. The probability cannot therefore be calculated.

It is possible, however, to calculate the average number of defects in a given unit length of cable or the average number of customers entering a shop in a given unit of time. Often it is found that the arrival of customers per unit of time or the number of defects per unit length (and similar cases where the variable varies per unit of time or length) have a Poisson distribution which is determined solely by a single parameter—the mean.

Example 14.

Number of defects per metre of cable	0	1	2	3	4	5	6
Number of metre lengths	14	26	30	17	5	5	3

On the assumption that the distribution of the number of defects per metre is a Poisson distribution what are the probabilities of their being 0, 1, 2, 3 . . . defects per metre?

Number of defects per metre of cable x	Number of metre lengths f	fx	Poisson probability $m = \dfrac{200}{100} = 2$		Expected frequency for a total frequency of 100
0	14	0	$P(0) = e^{-2}$	$= 0.13534$	14
1	26	26	$P(1)$	$= 0.27068$	27
2	30	60	$P(2)$	$= 0.27068$	27
3	17	51	$P(3)$	$= 0.18045$	18
4	5	20	$P(4)$	$= 0.09022$	9

5	5	25	$P(5)$	= 0.03609	4
6	3	18	$P(6)$	= 0.01203	1
	100	200		0.99549	100

Comparison between the expected and actual frequencies would suggest that the distribution of defects was a Poisson distribution. However it would be necessary to test whether the differences arose by chance. The test used is the *chi-squared test* for goodness of fit (*see* Chapter 21).

QUESTIONS

1. The incidence of occupational disease in an industry is such that employees have a 25 per cent chance of suffering from it. What is the probability that out of 5 workmen chosen at random 3 or more will suffer from the disease?

2. It is expected that 10 per cent of the production from a continuous process will be defective and scrapped.
Determine the probability that in a sample of 10 units chosen at random:
(a) exactly two will be defective; and
(b) at the most two will be defective; using in each case both (i) the binomial distribution; and (ii) the Poisson approximation to the binomial distribution.
Institute of Cost and Management Accountants.

3. One third of the sweets manufactured for packets of "Lucky Five" were pink. Each packet contains five sweets. Out of 243 packets how many would you expect to contain at least two pink sweets?

4. The output of an automatic machine shows that one in five pieces must be rejected. In a random sample of 8 pieces what is the probability of:
(a) the first three being accepted and the next four rejected?
(b) the first being accepted and then alternate rejections and acceptances?
(c) exactly 3 rejections?
(d) less than 3 rejections?
(e) at most 5 rejections?
Are you more likely to find no rejections than some rejections?

5. Robin Hood hits the target three times out of four. What is his probability of hitting the target three times in the next six shots?

6. Three drivers out of ten were wearing safety belts. There is

a pile-up of eight cars. What is the probability that less than half the drivers were wearing safety belts?

7. An item is made in three stages. At the first stage it is formed on one of four machines, A, B, C and D, with equal probability. At the second stage it is trimmed on one of three machines, E, F and G, with equal probability. Finally, it is polished on one of two polishers, H and I, and is twice as likely to be polished on the former, as this machine works twice as quickly as the other.

(a) What is the probability that an item is:
    (i) polished on H?
    (ii) trimmed on either F or G?
    (iii) formed on either A or B, trimmed on F and polished on H?
    (iv) either formed on A and polished on I or formed on B and polished on H?
    (v) either formed on A or trimmed on F?

(b) Suppose that items trimmed on E or F are susceptible to a particular defect. The defect rates on these machines are 10 per cent and 20 per cent respectively. What is the probability that an item found to have this defect was trimmed on F?

Association of Certified Accountants.

8. (a) State the addition, multiplication and conditional laws of probability.

(b) Out of a production batch of 10,000 items, 600 are found to be of wrong dimensions. Of the 600 which do not meet the specifications for measurement, 200 are too small. What is the probability that if one item is taken at random from the 10,000.

    (i) it will be too small;
    (ii) it will be too large?

If two items are chosen at random from the 10,000, what is the probability that:

    (i) both are too large;
    (ii) one is too small and one too large?

Institute of Chartered Secretaries and Administrators.

9. (a) Candidates' marks in an examination in Statistics were found to have a mean of 50 per cent and a standard deviation of 10 per cent. The marks were normally distributed. What proportion of candidates had marks (i) above 70 per cent (ii) below 40 per cent?

(b) If the marks gained referred to the results of 100 candidates, what is the reliability of the mean mark of 50 per cent (use 95 per cent confidence limits)?

Table of Areas of the Normal Curve
Standardised deviate   Area to the left of the standardised deviate

Standardised deviate	Area to the left of the standardised deviate
1	0.8413
1.96	0.9750
2	0.9772
3	0.9987

Institute of Chartered Secretaries and Administrators.

CHAPTER 20

Sampling and Significance

SAMPLE AND POPULATION

A *population* is an aggregate of similar things. A *sample* is a part of this population and must not be drawn from any other. The statistician is usually faced with data obtained from a sample and has to draw conclusions about the population from which the sample is taken; from the sample statistic he has to estimate the population parameter.

LARGE AND SMALL SAMPLES

There is a difference in treatment between large and small samples. A sample is considered large if it has at least 30 items.

THE THREE DISTRIBUTIONS—POPULATION, SAMPLE AND SAMPLE STATISTIC

Note. It is extremely important not to confuse these three distributions and to realise fully the relationship between them.

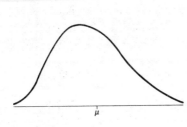

Fig. 77. Distribution of population values.

Figure 77 shows the *distribution of the individual values of the population.* Its mean is denoted by μ (pronounced mu it is the Greek letter "m").

Figure 78 shows the *distribution of the individual values of a single sample* of size n from this population. m denotes the mean value of this sample. It is extremely unlikely to be the same as μ.

182

Figure 79 shows the *distribution* of the *means of all possible samples* of size *n* which can be drawn from this population. m_s denotes the mean of one sample (the means vary from sample to sample). The distribution of the sample means is a "normal" distribution even though the population and a single sample may not be "normally" distributed *provided (a)* the samples are large samples and *(b)* the population is not too skewed.

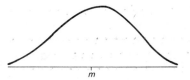

Fig. 78. Distribution of values of a single sample.

The mean value of the sample means will be the population mean.

Greek letters denote population parameters and roman letters sample statistics.

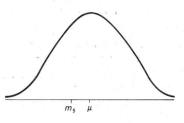

Fig. 79. Distribution of sample means.

STANDARD ERROR OF THE MEAN

The differences between the sample means and the population mean are *sampling errors*. The standard deviation of the sample means is therefore known as the standard error of the mean.

The standard error of the mean (for large samples) $= \dfrac{\sigma}{\sqrt{n}}$ where σ is the standard deviation of the population and *n* is the number of values in the sample.

Example 1. The Excel Tea Company sells packets of tea whose mean weight is 100 grams and whose standard deviation is 15 grams. Calculate the standard error of the mean for random samples of 36 packets.

$$\text{s.e.}_{\text{mean}} = \frac{\sigma}{\sqrt{n}} = \frac{15}{\sqrt{36}} = 2.5 \text{ grams.}$$

Where the standard deviation of the population is not known, as is usually the case, the standard deviation of the sample (provided it is a large sample) is a good enough estimate. The standard error of the mean then becomes $\dfrac{s}{\sqrt{n}}$ (*s* denoting the standard deviation of the sample).

Example 2. Bestham Retail Stores purchased from Excel Tea Company (*see* Example 1) a large consignment of 100 gram packets of tea. A random sample of 36 packets from this consignment had a mean weight of 99.5 grams and a standard deviation of 15.6 grams. Calculate the standard error of the mean.

Since the standard deviation of the population is not known the standard deviation of the sample must be used.

$$\text{s.e.}_{\text{mean}} = \frac{s}{\sqrt{n}} = \frac{15.6}{\sqrt{36}} = 2.6 \text{ grams.}$$

CONFIDENCE LIMITS

Given the mean value of a random sample, it is possible to give, with given probabilities of being right, limits within which the mean value of the population from which the sample was drawn, will lie. These limits are *confidence limits*.

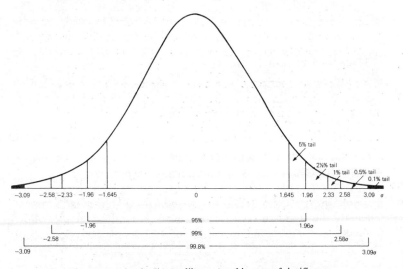

Fig. 80. Areas under "normal" curve used in tests of significance.

Since for large samples, the sample mean is "normally" distributed, it follows that 95 per cent of all possible sample means will have a value within a range of $\mu \pm 1.96$ s.e. (*see* Fig. 80). If then, it is assumed that the particular random sample drawn is one of the samples lying within this range, there will be a 95 per cent chance of being right in saying that μ lies between $m_s + 1.96$ s.e. and $m_s - 1.96$ s.e. This is known as the 95 per cent *confidence interval* and m_s + 1.96 s.e. and m_s − 1.96 s.e. are the *confidence limits*. Similarly m_s ± 2.58 s.e. gives the 99 per cent confidence limits (*see* Fig. 80). The confidence limits m_s + 3.09 s.e. and m_s − 3.09 s.e. means that there is only 1 chance in 1,000 that the population mean is less than the lower limit and 1 chance in 1,000 that it is greater than the upper limit (*see* Fig. 80).

Example 3. From the data given in Example 2 calculate the 95 per cent and 99 per cent confidence limits.
95% confidence limits:
        99.5 ± 1.96 (2.6), i.e. 94.4 grams and 104.6 grams.
99% confidence limits:
        99.5 ± 2.58 (2.6), i.e. 92.8 grams and 106.2 grams.

STANDARD ERROR OF A PROPORTION

Let Π (a Greek capital letter—pi) be the proportion of individuals in a population possessing a certain attribute (known as "successes"); then, provided the samples are large and Π is not too small, the proportion of "successes" in the samples will be "normally" distributed and the standard deviation, known as the standard error of a proportion, will be $\sqrt{\dfrac{\Pi(1-\Pi)}{n}}$. When Π is not known, as is usually the case, p the proportion of "successes" in the sample, is used instead. The standard error of a proportion then becomes

$$\sqrt{\frac{pq}{n}} \quad (q \text{ being } 1 - p).$$

The 95% confidence limits of a proportion are $p \pm 1.96\sqrt{\dfrac{pq}{n}}$.

The 99% confidence limits of a proportion are $p \pm 2.58\sqrt{\dfrac{pq}{n}}$.

Example 4. Of a random sample of 400 people interviewed, 80 used a certain brand of toothpaste. The manufacturer wishes to know what proportion of the whole population uses his brand of toothpaste.
Proportion of successes = $\dfrac{80}{400}$ = 0.20.

    $p = 0.20;$ $q = 1 - p = 0.80;$ $n = 400.$

$$\text{s.e.}_{\text{propn.}} = \sqrt{\frac{(0.20)(0.80)}{400}} = 2\%.$$

95% confidence limits = 20 ± (1.96)(2) = (20 ± 4)%.
99% confidence limits = 20 ± (2.58)(2) = (20 ± 5.2)%;

The maximum standard error of a proportion is when p = ½. If, therefore, the calculation is made as if this were the actual proportion there is a greater probability that the population proportion will be in the given range.

Example 5. Twenty per cent of the students of a college are taking accountancy courses. What is the chance of getting at least 24 such students in a random sample of 144?

Proportion of accountancy students in college = 0.20.

Proportion of non-accountancy students = 1 − 0.20 = 0.80.

Proportion of accountancy students in sample = $\dfrac{24}{144}$ = 0.17.

Difference between population and sample proportions
$$= 0.17 - 0.20 = - 0.03.$$

Standard error of proportion = $\sqrt{\dfrac{(0.20)(0.80)}{144}}$ = 0.03.

$Z = \dfrac{-0.03}{0.03} = -1$ s.e.

Area under "normal" curve from −1 s.e. to 0 s.e. = 0.3413 (*see* table in Appendix C).

Area under "normal" curve from −1 s.e. upwards = 0.3413 + 0.5 = 0.8413 (*see* Fig. 81).

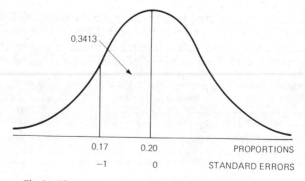

0.17	0.20	PROPORTIONS
−1	0	STANDARD ERRORS

Fig. 81. Diagram as an aid to solving "normal" curve problems.

Probability of getting at least 24 (i.e. 24 or more) students taking accountancy courses in a random sample of 144—proportion of 0.17 or more—is 0.84 or 84%.

Note. Many students have found diagrams of the type shown in Fig. 81 very helpful in solving problems involving the "normal" curve. A template made of cardboard is a useful device, especially for those not proficient at freehand drawing.

SIGNIFICANCE

If an actual observation is compared with some theoretical expectation there will usually be a difference. This difference may be due to the fluctuations of random sampling (sampling error), or the difference may be significant, that the theoretical expectation is wrong, i.e. the *null hypothesis* that there should be no difference is to be rejected. The problem is to determine whether the difference is significant. This can only be done in terms of probabilities.

SIGNIFICANCE OF THE MEAN (LARGE SAMPLES)

It has been shown that for large samples the probability of the population mean lying between $m_s \pm 1.96$ s.e. was 95 per cent; it follows that the probability of the population mean (μ) being outside this interval is 5 per cent; that is equivalent to saying that the probability of $\dfrac{m_s - \mu}{\text{s.e.}_{\text{mean}}}$ being greater than 1.96 or less than -1.96 is 5 per cent. A difference outside these limits is said to be *significant at the 5 per cent level.* There is only a 5 per cent chance of the difference being a sampling difference; the null hypothesis is rejected and the sample is considered as not being from the population whose mean value is μ. There is a 5 per cent probability of being wrong in rejecting the null hypothesis. If it is required to take a smaller chance of being wrong then a *1 per cent level of significance* could be chosen which would require $\dfrac{m_s - \mu}{\text{s.e.}_{\text{mean}}}$ to be greater than 2.58 or less than -2.58. Since the probability of the population mean being outside this range would be only 1 per cent, rejecting the null hypothesis would lead to only 1 per cent probability of being wrong.

Example 6. A sample of 3,244 households had a mean size of 3.11 persons and a standard deviation of 1.63 persons. Test whether the mean size of households has changed since a census of population gave a mean size of 3.04 persons.

The null hypothesis is that there has been no change (that the difference in mean size of household is a sampling difference).

$$\text{Standard error of mean} = \frac{1.63}{\sqrt{3,244}} = 0.03.$$

$$Z = \frac{m_s - \mu}{\text{s.e.}_{\text{mean}}} = \frac{3.11 - 3.04}{0.03} = 2.3.$$

This is significant at the 5 per cent level but not at the 1 per cent level. If the decision is to be made at the 5 per cent level of significance the null hypothesis is rejected and the conclusion drawn that the mean size of households has changed since the census. If, however, the decision is to be made at the 1 per cent level, the null hypothesis would be accepted and the mean size of households would be considered unchanged.

TWO TAIL OR ONE TAIL TEST

In Example 6 it was a question of whether the size of the household had changed, not whether it was bigger or smaller. The sample mean could have been smaller than the population mean as well as, as in this case, bigger. It was therefore a question of finding the *probability of a difference* occurring irrespective of whether the sample mean was larger or smaller than the population mean. On referring to Fig. 80 it will be seen that the *probability of a difference* is denoted by the *area of both tails.*

On the other hand when it is a question of the *probability* of the sample mean being *either* larger or smaller than the population mean, the probability will be denoted by the area of *one tail.*

Example 7. The Barbalonia Drug Company claims that the average contents of their bottles of hair restorer is 100 cc. A random sample of 144 bottles was found to have an average amount of 99 cc per bottle with a standard deviation of 4 cc. Should the company's claim be rejected or accepted at the 0.005 level of significance?

The point of concern is that the bottles should not contain less than 100 cc. The null hypothesis is that the bottles do not contain less than 100 cc.

$$\text{s.e.}_{\text{mean}} = \sqrt{\frac{4}{144}} = 0.33 \text{ cc}; \qquad Z = \frac{99 - 100}{0.33} = -3.$$

Critical value of 0.005 level of significance (1 tail) is 2.58. The company's claim must be rejected.

Critical values of Z

Level of significance	0.10	0.05	0.01	0.005	0.002
One tail test	1.28	1.645	2.33	2.58	2.88
Two tail test	1.645	1.96	2.58	2.81	3.09

Level of significance	*How regarded*
0.10	not significant
0.05	significant
0.01	
0.005	highly significant
0.002	very highly significant

SIGNIFICANCE OF THE DIFFERENCE BETWEEN MEANS (LARGE SAMPLES)

It often happens that samples are from two different sources and it is required to know whether the difference in their means arises from their being a significant difference (that is, they are samples from different populations) or if the difference is a sampling difference (that is, the samples, although from different sources, really belong to the same population).

Where the samples are large (30 or more), the distribution of the difference between their means is a "normal" distribution, provided the samples are chosen quite independently of each other.

The standard error of the difference between the means of two samples is:

$$\sqrt{\frac{s_1{}^2}{n_1} + \frac{s_2{}^2}{n_2}}$$

where s_1 is the standard deviation of sample 1 and s_2 is the standard deviation of sample 2—the standard deviation of large samples being estimates for the standard deviations of the populations—and n_1 is the number of values in sample 1 and n_2 is the number of values in sample 2.

It is convenient to use the term *variance* for the square of the standard deviation; thus $s_1{}^2$ is referred to as the variance of sample 1.

The Z statistic for the difference between means of large samples is:

$$\frac{m_1 - m_2}{\text{s.e.}_{\text{diff. mean}}}$$

where m_1 is the mean of sample 1 and m_2 is the mean of sample 2.

Example 8. Two manufacturers Alpha and Omega supplied steel rods (nominal length 1,200 mm). A sample of 100 rods supplied by Alpha had a mean length of 1,190 mm and a standard deviation of 90 mm. A sample of 75 rods supplied by Omega had a mean length of 1,230 mm and a standard deviation of 120 mm. Is there a difference in the length of the rods supplied by these two manufacturers? Test at *(a)* the 5 per cent level of significance, *(b)* the 1 per cent level.

Null hypothesis: there is no difference in the length of rods from the two manufacturers.

Testing for a *difference,* therefore two tail test.

$$\text{s.e.}_{\text{diff. mean}} = \sqrt{\frac{90^2}{100} + \frac{120^2}{75}} = 16.52 \text{ mm.}$$

$$Z = \frac{1,190 - 1,230}{16.52} = -2.42.$$

(a) Significant at 5 per cent level. (Critical value of Z is 1.96). Null hypothesis rejected: the rods are of different lengths.

(b) Not significant at 1 per cent level. (Critical value of Z is 2.58.) Null hypothesis accepted: there is no difference in the lengths of rods.

Example 9. Given the same data as in Example 8, test the hypothesis that the rods of manufacturer Omega are longer than those of manufacturer Alpha. Test at *(a)* the 5 per cent level, *(b)* the 1 per cent level. Are the results contradictory to those of Example 8?

Null hypothesis: The rods manufactured by Omega are not longer.

Testing that the rods from Omega are *longer,* therefore one tail test.

$$Z = 2.42.$$

(a) Significant at 5 per cent level. (Critical value of Z is 1.645.) Null hypothesis is rejected: Omega's rods are longer.

(b) Significant at 1 per cent level. (Critical value of Z is 2.33.) Null hypothesis is rejected: Omega's rods are longer.

In Example 8 the difference in length was not significant at the 1 per cent level whereas in Example 9 the longer length was significant at this level. This may appear contradictory. But 1 per cent for one tail is *not* the same probability as 1 per cent for two tails. Reference to Fig. 80 should make this clear. One per cent probability that the rods are longer than a Z value of 2.33 implies a 1 per cent probability that the rods are shorter than a Z value of -2.33; there is therefore a 2 per cent probability that the length of the rods differ by more than a Z value of 2.33.

The results of Examples 8 and 9 are therefore consistent.

PROCEDURE

Testing the significance of the mean or the difference between means (large samples)

1. Determine the null hypothesis.
2. Determine whether one tail or two tail test is required.
3. Compute standard error.
4. Compute Z statistic.
5. Choose level of significance and hence the critical value of Z.
6. If Z is greater than the critical value, reject the null hypothesis; if less, accept the null hypothesis.

THE SIGNIFICANCE OF A PROPORTION

It has already been shown that for large samples where the proportion of items possessing a given attribute in a population is not too small (pn and qn must be greater than 5), the sample proportions are distributed "normally".

Example 10. A manufacturer claimed that at least 96 per cent of the equipment he supplied was in conformity with the specifications. An examination of a sample of 200 pieces of equipment showed that 14 pieces were not in accordance with the specifications. Test his claim at the *(a)* 1 per cent, and *(b)* 5 per cent levels of significance.

$$n = 200; \qquad p = 0.04; \qquad q = 0.96; \qquad pn = 8;$$
"normal" approximation appropriate.
(p = proportion of defective equipment).

Null hypothesis: in conformity with specification.

Testing for *different* specifications, therefore two tail test.

$$\text{s.e.}_{\text{propn.}} = \sqrt{\frac{(0.04)(0.96)}{200}} = 0.01385.$$

Proportion of defective equipment in sample: $\frac{14}{200} = 0.07$.

$$Z = \frac{0.07 - 0.04}{0.01385} = 2.17.$$

(a) Not significant at 1 per cent level: claim upheld.
(b) Significant at 5 per cent level: claim rejected.

Example 11. A sample of 100 households showed 13 per cent to be living below a defined poverty line. Does this indicate an improvement over a previously established figure of 15 per cent?

pn and qn greater than 5: "normal" approximation appropriate.

Null hypothesis: no improvement.

Testing for an improvement (*not* a change — either an improvement or a deterioration), therefore a one tail test.

$$\text{s.e.}_\text{propn.} = \sqrt{\frac{(0.15)(0.85)}{100}} = 0.036.$$

$$Z = \frac{0.15 - 0.13}{0.036} = 0.6$$

Null hypothesis accepted: there is no indication of an improvement.

SIGNIFICANCE OF A DIFFERENCE BETWEEN PROPORTIONS

Consider two populations. From one of them is drawn a *large* sample (size n_1) of which there is a proportion p_1 having a given attribute. From the other population a *large* sample (size n_2) is also drawn of which there is a proportion p_2 having the given attribute. Provided that $p_1 n_1$, $q_1 n_1$, $p_2 n_2$ and $q_2 n_2$ are all greater than 5, the distribution of the differences between the proportions will be a "normal" distribution.

The null hypothesis assumes that the samples come from the same population and that the differences in the proportions between the samples are sampling differences. On this assumption the best estimate of the population proportion is $\dfrac{p_1 n_1 + p_2 n_2}{n_1 + n_2}$ denoted by p.

The standard error of the difference between proportions is

$$\sqrt{pq\left(\frac{1}{n_1} + \frac{1}{n_2}\right)}; \qquad Z = \frac{p_1 - p_2}{\text{s.e.}_\text{diff. propn.}}$$

Example 12. A certain product was manufactured by two different processes, A and B. Of a random sample of 400 pieces made by process A, 100 were of second quality. Of a random sample of 200 pieces made by B, 44 were of second quality. Test at the 5 per cent level whether process B is better than process A.

pn and *qn* are greater than 5 in respect of each sample; "normal" approximation appropriate.

Null hypothesis: there is no difference.

Testing the hypothesis that process B is better than process A, therefore one tail test.

Estimation of population proportion $= \dfrac{100 + 44}{400 + 200} = 0.24.$

$$p = 0.24; \qquad q = 1 - 0.24 = 0.76;$$

$$p_1 = \frac{100}{400} = 0.25; \qquad p_2 = \frac{44}{200} = 0.22.$$

$$\text{s.e.}_{\text{diff. propn.}} = \sqrt{(0.24)(0.76)\left(\frac{1}{400} + \frac{1}{200}\right)} = 0.037.$$

$$Z = \frac{0.25 - 0.22}{0.037} = 0.81.$$

Not significant at 5 per cent level: null hypothesis accepted—process B is not better than process A.

PROCEDURE

Testing the significance of a proportion or the difference between proportions (large samples where pn and qn are greater than 5)

1. Ensure that *pn* and *qn* are greater than 5.
2. Establish the null hypothesis.
3. Determine whether a one or two tail test is required.
4. If the population proportion is unknown, estimate from sample proportions.
5. Compute standard error.
6. Compute *Z* statistic.
7. Choose level of significance and hence the critical value of *Z*.
8. If *Z* is greater than the critical value, reject the null hypothesis; if less, accept the null hypothesis.

SMALL SAMPLES

When a sample is small, that is less than 30 items, the standard deviation of the sample can no longer be regarded as an estimate of the standard deviation of the population. The estimate of the standard deviation of the population for small samples is obtained by multiplying the standard deviation of the sample by $\sqrt{n/(n-1)}$ (known as Bessel's correction) where n is the number of items in the sample. Another way of arriving at this estimate is from the values of the sample: divide $\Sigma(x - \bar{x})^2$ by $n - 1$ instead of n to obtain an estimate of the population variance. The standard deviation is the square root of this. This estimate is known as the *unbiased estimate.*

This unbiased estimate (obtained by dividing the " sum of the squares" by $n - 1$) is the correct method in all cases but when n is large, the difference is negligible.

Example 13. Five ladies followed a special slimming diet and after two weeks showed a loss of weight of:

$$4, 6, 3, 5, 2 \text{ kg.}$$

Estimate the standard deviation of the population from which this sample could have come.

x	x^2
4	16
6	36
3	9
5	25
2	4
20	90

$n = 5$
$\Sigma x = 20$
$\Sigma x^2 = 90$
$\bar{x} = \dfrac{20}{5} = 4$

Sample variance $\left(s^2 = \dfrac{\Sigma x^2 - n\bar{x}^2}{n}\right)$

$$s^2 = \frac{90 - 5(16)}{5} = 2;$$

$$s = \sqrt{2} = 1.4 \text{ kg.}$$

Population variance $\left(\sigma^2 = \dfrac{\Sigma x^2 - n\bar{x}^2}{n-1}\right)$

$$\sigma^2 = \frac{90 - 5(16)}{4} = 2.5;$$

$$\sigma = \sqrt{2.5} = 1.6 \text{ kg.}$$

Note. The "sum of the squares" is $\Sigma(x - \bar{x})^2 = \Sigma x^2 - n\bar{x}^2$. Divided by n this will give the variance of the sample; divided by $n - 1$, it gives the estimated variance of the population.

Given the standard deviation of the sample, the estimated standard deviation of the population is $s(\sqrt{n/(n-1)})$.

In Example 13, $s = \sqrt{2}$; $\sigma = (\sqrt{2})(\sqrt{5/4}) = 1.6$ (as before). When samples are small, the means are *not* "normally" distributed. The t distribution (*see* below) must be used. The t distribution does, however, assume that the population is "normally" distributed.

THE t DISTRIBUTION

This distribution is symmetrical around the mean (which coincides with the mode and median) and like the normal distribution, the variable ranges from minus infinity to infinity.

There is a t curve for each number of the *degrees of freedom* (denoted by ν—the Greek letter "nu" or by the letters d.f.—*see* below). The greater ν is, the more the t curve approaches the "normal" curve.

The areas under the t curve, as in the case of the "normal curve", represent probabilities.

Appendix D gives only part of a t table. A more complete table would give other probabilities and other numbers of degrees of freedom. The areas given are for two tails.

DEGREES OF FREEDOM

In computing certain statistics, such as the t statistic, the sample observations together with certain population parameters are used. When the parameters are unknown they have to be estimated from the sample. The number of degrees of freedom is the number of independent observations *less* the number of population parameters that have to be estimated from the sample. In Example 13 the number of degrees of freedom is 5—the number of independent observations—*less* 1—the standard deviation of the population—making the number of degrees of freedom 4.

THE MEAN OF A SINGLE SAMPLE (SMALL)

Example 14. Was the loss of weight of the 5 ladies who were slimming in Example 13 significant?

Null hypothesis: no difference in weight after the diet; the population mean is zero.

Testing for loss, therefore one tail test.

$$\text{s.e.}_{\text{mean}} = \frac{\sigma}{\sqrt{n}} = \sqrt{\frac{1.6}{5}} = 0.71 \text{ kg.}$$

$$t = \frac{4-0}{0.71} = 5.6.$$

Number of degrees of freedom $= n - 1 = 5 - 1 = 4$.

From the table in Appendix D a t value of 5.60 or more for 4 degrees of freedom has a probability of 0.01 for two tails, that is a probability of 0.005 for one tail.

The loss of weight is significant at the 0.005 level — strong statistical evidence that the diet was effective.

Note. When the standard deviation was computed from the individual values of the sample as in Example 13, students sometimes take this to be the standard error of the mean which, of course, requires a further calculation.

DIFFERENCE BETWEEN MEANS OF SMALL SAMPLES

The null hypothesis that there is no difference between the samples (other than sampling differences) and the necessity to estimate the population standard deviation from the samples standard deviations postulate that the sampling distribution of the differences between the means is a t distribution with $n_1 + n_2 - 2$ degrees of freedom, having a mean equal to zero and a standard error of:

$$\sqrt{\left(\frac{n_1 s_1^2 + n_2 s_2^2}{n_1 + n_2 - 2}\right)\left(\frac{n_1 + n_2}{n_1 n_2}\right)}$$

Example 15. Concrete beams of the same type, made at two different works, were tested for strength with the following results.

		Works No. 1		Works No. 2
Number in sample	(n_1)	12	(n_2)	10
Mean strength kg/cm^2	(x_1)	5,000	(x_2)	4,975
Standard deviation kg/cm^2	(s_1)	50	(s_2)	60

Are the beams from works no. 1 stronger than those from works no. 2? Test at the 5 per cent level of significance.

Null hypothesis: not stronger.

Testing whether beams from works no. 1 are stronger, therefore one tail test.

$$\text{s.e.}_{\text{diff. mean}} = \sqrt{\left(\frac{(12)(50)^2 + (10)(60)^2}{12 + 10 - 2}\right)\left(\frac{12 + 10}{12 \times 10}\right)} = 24.59 \text{ kg/cm}^2 .$$

$$t = \frac{5,000 - 4,975}{24.59} = 1.02.$$

$$\text{d.f.} = 12 + 10 - 2 = 20.$$

The t table in Appendix D gives values for two tails. Since the value of the t statistic which denotes a 5 per cent probability for one tail also denotes a 10 per cent probability for two tails, it will be necessary to find the critical value for 10 per cent with 20 d.f. This is 1.725.

The null hypothesis is accepted; the beams from works no. 1 are not stronger.

PROCEDURE

Testing the significance of the mean and the difference between means of small samples

1. Determine the null hypothesis.
2. Determine whether one or two tail test is required.
3. Compute standard error.
4. Compute t statistic.
5. Compute the number of degrees of freedom; $(n - 1)$ for a single mean, $(n_1 + n_2 - 2)$ for a difference between means.
6. Choose level of significance and hence critical value of t.
7. If t greater than critical value, reject null hypothesis; if less, accept null hypothesis.

QUESTIONS

1. Define simple random sampling. A random sample of 1,000 entries in a set of accounts includes 35 incorrect entries. What is the range of the proportion of defective entries in the complete population of entries in the accounts? (Use limits of ± 2 standard errors of the proportion.) In a check of 2,000 entries in a different set of accounts, 90 were found to be in error. Is there any significant difference between the two proportions of errors in entry in the two different sets of accounts?

The Institute of Chartered Secretaries and Administrators.

2. *(a)* The mean value of accounts at least three months in arrear for a set of 500 company accounts is £400. If the standard deviation of accounts in arrear is £100, what is the reliability of this mean?

(b) A second company finds that its mean value of accounts three months in arrear or more is £380. The number of accounts involved is 750, and the standard deviation of accounts in arrear is £60. Are the two means significantly different?

The Institute of Chartered Secretaries and Administrators.

3. *(a)*

(i) Define the standard error of a mean of samples of n observations and use the concept to explain briefly why one has more confidence in the mean of a large sample than in the mean of a small one.

(ii) Show why it is unlikely that a sample of 64 observations

with a mean of 5.2 was drawn from a population with a mean of 5.5 and a standard deviation of 0.8.

(b) An expensive piece of equipment is known to be utilised for about 20 per cent of the time. In order to obtain a more accurate assessment of its utilisation the management decides to carry out some activity sampling, by observing the equipment at random instants of time and noting the percentage of occasions on which it is working.

Determine how many observations will be required to be able to state the percentage utilisation within ± 5 per cent with 95 per cent certainty.   *The Association of Certified Accountants.*

4. (a) A mortgage manager of a building society has taken the view in recent years that 10 per cent of mortgage applications received will be unacceptable. 1,000 applications received at the end of 19 . . were investigated and 125 were found to be unacceptable. Does this sample evidence appear to be compatible with the manager's experience, or is there a significant difference between the sample proportion and experience?

(b) A second building society took a sample of 500 application forms at the same time as that used for the basis of the sample in (a) above. It found 77 were unacceptable. Is there a significant difference between acceptability ratios for the two societies?
        Institute of Chartered Secretaries and Administrators.

5. In order to test the effectiveness of a drying agent in paint, the following experiment was carried out. Each of six samples of material was cut into two halves. One half of each was covered with paint containing the agent and the other half with paint without the agent. Then all twelve halves were left to dry. The time taken to dry was as follows:

Drying time (hours)					
sample number					
1	2	3	4	5	6
Paint with agent 3.4	3.8	4.2	4.1	3.5	4.7
Paint without agent 3.6	3.8	4.3	4.3	3.6	4.7

Carry out a t test to determine whether the drying agent is effective, giving your reasons for choosing a one tail or a two tail test. Carefully explain your conclusions.
                The Association of Certified Accountants.

6. The following represents a random sample of daily sales (in £s) of a trader and refers to one line only:

120, 98, 114, 80, 131, 118, 130, 110, 105, 107.

Calculate his average daily sales of this line and the standard deviation.

If he could not replenish until the following morning, what value of stock should be hold to be 95 per cent sure that he would not run out that day? (Assume that the distribution is normal.)

Institute of Cost and Management Accountants.

7. The mean breaking strength of steel rods is given as 25,000 kg. The breaking strength of 120 rods were found to have a mean of 23,800 kg with a standard deviation of 560 kg. Is the complaint that the rods were not up to specification justified?

If only 60 rods had been tested and were found to have the same mean and standard deviation would the complaint still be justified?

8. A machine making components to a nominal dimension of 3.000 cm is reset every morning. The first 36 components produced one morning have a mean of 2.988 cm with a standard deviation of 0.004 cm. Does this provide sufficient evidence that the machine is set too low?

9. A machine produces metal parts and feeds them into a bin. An empty bin is placed by the machine after every 100 parts and the full bin wheeled away for inspection. After checking a large number of bins the inspector noted that per bin of 100 parts the average number of defective parts was 10 and the standard deviation of the number of defective parts was 3.

Assuming that production continues under the same conditions:

(a) if larger bins were used each holding 300 parts what would be:

(i) the average number of defective parts per bin.

(ii) the standard deviation of the number of defective parts per bin?

(b) how many parts must the bin be able to hold so that the standard deviation of the number of defective parts per bin is equal to 1 per cent of the total number of parts in the bin?

Assuming that owing to a change in the conditions of production the machine produces good and bad parts in random order in the proportion of 99 : 1 and that bins each holding 100 parts are used:

(c) estimate the proportion of bins which would contain no defective parts, and

(d) what would you expect the standard deviation of the number of defective parts to be?

Show how you arrive at your answers.

Institute of Cost and Management Accountants.

10. The manufacturers of brand "X" margarine held a tasting test to determine whether people could distinguish between their product and butter. Each subject was presented with 6 biscuits, 5 of

which were spread with butter and one with brand "X" margarine, and asked to pick out the margarine.

Of 1,680 people tested, 345 picked out the margarine-spread biscuit correctly.

(a) What proportion of people would you expect to pick out the odd biscuit, if their choice were purely random?

(b) Could the results given above be purely due to chance or could some people tell brand "X" margarine from butter?

(c) How does the theory of the normal curve help in assessing the results?   *Institute of Cost and Management Accountants.*

11. A chocolate manufacturer makes bars of chocolate which have a mean weight of 115 grams and a standard deviation of 12 grams. The manufacturer advertises that the minimum weight is 105 grams. What proportion of bars is likely to be less than this advertised minimum weight?

The manufacturer subsequently decides to advertise that the minimum weight of a bar will be raised to 110 grams. If the standard deviation remains at 12 grams, what mean weight must be obtained to make sure that only one per cent of the bars are less than the advertised minimum weight of 110 grams? Assume that the weights of all bars are distributed "normally".

12. What is meant by the term "the standard error of a proportion observed from a random sample"? In a recent survey, a random sample of 10,000 adults provided 62 per cent who thought that cigarette smoking affected health. Calculate the standard error of this proportion and thus indicate the percentage of adults in the population who took this view, showing a range with 95 per cent confidence limits. What sample size would have been necessary to halve these limits?

13. In a survey of customers' accounts, the following grouped frequency table was obtained dealing with credit made available to clients (in weeks):

Credit Available to Customers

Credit period provided (weeks)	Number of customers
3	46
4	73
5	117
6	35
7	12
8	7

What is the mean credit period provided? In a similar firm it is found that the mean credit period was 5½ weeks, (the standard deviation was 1 week). Is there any significant difference between the mean credit periods provided by the two firms?

14. What is the importance of the "normal" distribution in statistical analysis?

A manufacturer produces tyres which are expected to have a mean life of 24,000 miles. (The statistics of tyre life are "normally" distributed.) If the standard deviation of tyre life is 1,000 miles:

(a) what proportion will have worn out after being used for 22,000 miles?

(b) what proportion will have been worn out after 25,000 miles of use?

What answers would you have obtained if the mean life had been raised to 24,500 miles?

The London Chamber of Commerce and Industry.

CHAPTER 21

Chi-Squared

TESTS OF ASSOCIATION

CONTINGENCY TABLES

A contingency table consisting of m rows and n columns is an $m \times n$ contingency table. An example of a 3 × 2 table is shown in Fig. 82.

Political Affiliation of Employees

	Manual employment	Non-manual employment	Totals
Conservative	18	19	37
Labour	46	14	60
Liberal	11	13	24
Totals	75	46	121

(Hypothetical data)

Fig. 82. An example of a 3 × 2 table.

Each observation is classified in two ways; according to *(a)* political party, *(b)* type of employment. The above table shows, for example, that 46 out of a total of 121 observations consisted of people belonging to the Labour Party who were engaged in manual work.

The contingency table provides a means of testing if there is any association between the characteristics upon which the classification is based; in the above example, if there is any association between political party and type of work.

THE "NULL" HYPOTHESIS

This is the hypothesis that there is no relationship between the characteristics. In the above example, the null hypothesis is that there is *no association* between political party and type of work. If this hypothesis is correct, the proportion of Conservatives who are engaged in manual work will be the same as the proportion of Conservatives among all the observations; that is, the expected number of Conservatives engaged in manual work = 37/121 (the proportion of Conservatives in total) $\times$ 75 (the total engaged in manual employment) = 22.9

The other expected frequencies are obtained in a similar way. They are tabulated below:

	Manual employment	*Non-manual employment*	*Totals*
Conservative	22.9	14.1	37
Labour	37.2	22.8	60
Liberal	14.9	9.1	24
Totals	75.0	46.0	121

The totals must, of course, be the same as for the observed frequencies.

χ^2 (CHI-SQUARED)

This measures the difference between the expected frequencies and the observed frequencies, and is computed as follows:

$\chi^2 = \Sigma \dfrac{(O - E)^2}{E}$, where O is the observed frequency and E is the expected frequency in respect of each cell. In the above example

$$\chi^2 = \frac{(18 - 22.9)^2}{22.9} + \frac{(19 - 14.1)^2}{14.1} + \frac{(46 - 37.2)^2}{37.2} +$$

$$\frac{(14 - 22.8)^2}{22.8} + \frac{(11 - 14.9)^2}{14.9} + \frac{(13 - 9.1)^2}{9.1} = 10.9$$

DEGREES OF FREEDOM

The number of degrees of freedom for a contingency table $m \times n$ is $(m - 1)(n - 1)$. For the 3×2 table just given the number of degrees of freedom is $(3 - 1)(2 - 1) = 2 \times 1 = 2$.

Degrees of freedom are denoted by v (pronounced nu), n or by d.f.

The table above has 2 d.f. This means that only 2 expected frequencies need to be computed. The others are obtained by subtraction from marginal totals.

YATES'S CORRECTION

In the case of a 2 × 2 table there is only one degree of freedom, and it is necessary to make an adjustment in the computation of χ^2. *This adjustment is made only in the case of one degree of freedom.* This is done as follows:

If $(O - E)$ is, say, -3.2, then the adjusted $(O - E)$ is -2.7 (0.5 is deducted from 3.2) and $(O - E)^2$ is $(-2.7)^2$. If $(O - E)$ is, say, 3.2, then the adjusted $(O - E)$ is 2.7 (0.5 is again deducted from 3.2) and $(O - E)^2$ is 2.7^2. 0.5 is always taken from the absolute value of $(O - E)$ whether it is negative or positive.

PROCEDURE

Computing chi-squared as a test of association

1. Compute totals of columns and rows and the grand total of the observations (denoted by O).
2. Compute the expected frequencies (denoted by E) and ensure that the totals of rows, columns and the grand total are the same as those of the observed frequencies.
3. Compute the differences between observed and expected frequencies, i.e. $O - E$.

Note. In the case of a 2 × 2 table Yates's correction must be made.

4. Square the differences found in step 3 and divide by the expected frequencies: $\dfrac{(O - E)^2}{E}$.

5. Sum the values found in step 4: $\Sigma \dfrac{(O - E)^2}{E}$.

6. Determine the degrees of freedom.
7. From the degrees of freedom (step 6) and the value of chi-squared (step 5) determine the probability of association between the attributes.

THE SIGNIFICANCE OF CHI-SQUARED

If the null hypothesis is true, that is, there is no association between the two characteristics, yet there can still arise a difference between the expected frequencies and the observed frequencies. The greater χ^2 (which measures these differences), the smaller the probability of its occurring. The probability of χ^2 of any given amount also depends upon the number of degrees of freedom. In the case above, χ^2 was 10.9 with two degrees of freedom. Reference to tables of χ^2 (Appendix E) shows the probability of χ^2 being 10.9 or greater with two degrees of freedom is less than 0.5 per cent. This is such a small probability that the null hypothesis is rejected, and the evidence indicates there is a relationship between political party and type of employment.

GOODNESS-OF-FIT TESTS

Goodness-of-fit refers to how far an observed frequency distribution agrees with some theoretical distribution. In Example 14 of Chapter 19 a Poisson distribution was fitted to an observed distribution. The *differences* between the expected frequencies (on the hypothesis that the observed frequencies are a sample from a Poisson population) and the *observed frequencies* are tested to find the probability of the differences being sampling differences. The test used is the *chi-squared test.*

Example 1. The following data are taken from Example 14 of Chapter 19. Test whether the theoretical distribution is a good fit for the observed one.

Number of defects per metre	0	1	2	3	4	5	6
Number of metre lengths	14	26	30	17	5	5	3
Expected frequency	14	27	27	18	9	4	1

x	O	E	$O - E$	$\dfrac{(O - E)^2}{E}$
0	14	14	0	0.00
1	26	27	−1	0.04
2	30	27	3	0.33
3	17	18	−1	0.06
4	5	9	−4	1.78
5	5 \ 8	4 \ 5	3	1.80
6	3 /	1 /		
			$\chi^2 =$	4.01

Note. Since no class should contain less than 5 observations the 5-defects and 6-defects classes have been combined making a total of 6 classes.

In calculating the expected frequencies of the theoretical distribution the *mean* number of defects per metre and the *total* frequencies of the empirical distribution were used, hence there are two restrictions on the number of degrees of freedom.

$$\text{d.f.} = 6 - 2 = 4$$

With 4 d.f. the probability of obtaining x^2 of 4.01 or more is greater than 30 per cent (*see* Appendix E). The null hypothesis that the observed distribution is drawn from a population with a Poisson distribution is accepted.

Example 2. Several hundred women are employed on a certain assembly operation. The results of timing a sample of 100 women were as follows:

Time required to do job (minutes)	Number of women
20 and under 22	8
22 ” ” 24	24
24 ” ” 26	34
26 ” ” 28	28
28 ” ” 30	6
	100

Test the hypothesis that the distribution of time taken is a "normal" distribution.

x	mid-point	f	fx	fx^2
(20 - 22)	21	8	168	3,528
(22 - 24)	23	24	552	12,696
(24 - 26)	25	34	850	21,250
(26 - 28)	27	28	756	20,412
(28 - 30)	29	6	174	5,046
		100	2,500	62,932

$$\bar{x} = \frac{2,500}{100} \quad 25 \text{ minutes.}$$

$$s = \sqrt{\frac{62,932}{100} - 25^2}$$

$$= 2.08 \text{ minutes.}$$

$x - \bar{x}$	Z		Area under curve
$20 - 25 = -5$	$\dfrac{-5}{2.08} =$	-2.40	0.4918
$22 - 25 = -3$	$\dfrac{-3}{2.08} =$	-1.44	0.4251
$24 - 25 = -1$	$\dfrac{-1}{2.08} =$	-0.48	0.1844
$26 - 25 = \;\; 1$	$\dfrac{1}{2.08} =$	0.48	0.1844
$28 - 25 = \;\; 3$	$\dfrac{3}{2.08} =$	1.44	0.4251
$30 - 25 = \;\; 5$	$\dfrac{5}{2.08} =$	2.40	0.4918

Class interval	Area under curve	Expected frequency (area × total frequency)
under 20	$0.5000 - 0.4918 = 0.0082$	0.82 ⎱ 7.49
20 and under 22	$0.4918 - 0.4251 = 0.0667$	6.67 ⎰
22 ” ” 24	$0.4251 - 0.1844 = 0.2407$	24.07
24 ” ” 26	$0.1844 + 0.1844 = 0.3688$	36.88
26 ” ” 28	$0.4251 - 0.1844 = 0.2407$	24.07
28 ” ” 30	$0.4918 - 0.4251 = 0.0667$	6.67 ⎱ 7.49
30 and over	$0.5000 - 0.4918 = 0.0082$	0.82 ⎰

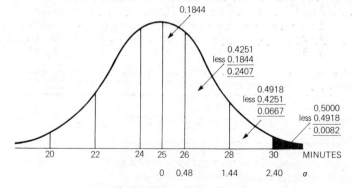

Fig. 83. Fitting a "normal" curve.

Time taken (minutes)	Observed frequencies	Expected frequencies	$O - E$	$\dfrac{(O - E)^2}{E}$
under 22	8	7.49	0.51	0.035
22 and under 24	24	24.07	−0.07	0.000
24 ″ ″ 26	34	36.88	−2.88	0.225
26 ″ ″ 28	28	24.07	3.93	0.642
28 and over	6	7.49	−1.49	0.296
			χ^2 =	1.198

In obtaining the theoretical expected frequencies, the mean, the standard deviation and the total frequency of the observed distributions were used, hence there are three restrictions on the number of degrees of freedom. There are five classes.

$$\text{d.f.} = 5 - 3 = 2$$

The probability of obtaining a χ^2 of 1.198 or more with 2 d.f. is over 30 per cent. The hypothesis that the time taken is "normally" distributed is accepted.

PROCEDURE

Testing goodness-of-fit

1. Choose the distribution to which the data is to be fitted, e.g. Poisson, "normal".
2. From the data obtain the values of the parameters of the distribution, e.g. mean, standard deviation.
3. From the values of these parameters compute the expected values.
4. Compute chi-squared.
5. Determine the number of degrees of freedom.
6. From the value of chi-squared and the number of degrees of freedom determine the goodness-of-fit.

QUESTIONS

1. Relationship between aggressive mood and illness behaviour in a sample of male employees in a large company.

| Aggressive mood | Frequency of illness behaviour | | | |
	Rare	Intermediate	Frequent	Total
Low	59	51	12	122
High	59	82	39	180
Total	118	133	51	302

Source: S.V. Kasl, *Journal of Chronic Diseases,* 1964.

On the assumption that the data are derived from a random sample, test the hypothesis that there is no association between aggressive mood and frequency of illness behaviour. Explain fully the meaning of your result.

B.Sc. (Sociology), University of London.

2. Explain briefly the purpose of the χ^2 (Chi-squared) test of association for contingency tables.

Change in Turnout of Electors and Swing to Labour in 95 Conservative non-rural Constituencies at the 1964 General Election

| Change in turnout | Swing | | Total |
	4% or less	more than 4%	
2.1% fall or more	10	27	37
Less than 2.1% fall, or rise	29	29	58
Total	39	56	95

Source: H.B. Berrington,
Journal of the Royal Statistical Society, 1965.

For the above data, test whether there is a significant association between change in turnout and swing to Labour. Interpret your result.   *B.Sc. (Sociology), University of London.*

3. The Acca Co. Ltd. tests all components built into its products to ensure that they have attained the standard of quality required. There are three checkers, A, B and C, who undertake this work. The table below shows the numbers of components accepted and rejected by the checkers when they tested three batches of component 1703 recently delivered to the factory. You are required to test the hypo-

thesis that the proportions of components rejected by the three are equal and explain the significance of your conclusion.

Use the chi-squared (χ^2) distribution as a test of significance, at the 0.05 level.

The Acca Co. Ltd.
Components quality test

	Checker A	Checker B	Checker C	Total
Accepted	44	56	50	150
Rejected	16	24	10	50

Association of Certified Accountants.

4. A petrol retailing organisation is concerned about the build up of queues at some of its busiest filling stations and consequent loss of business to competitors. A survey is carried out at selected stations to determine the arrival characteristics of customers. It is reasoned that by gaining this information the organisation will be better able to plan the layout of its forecourts in future.

At one selected filling station in a problem area the distribution of arriving customers was as follows:

No. of arrivals per 5-minute period	No. of 5-minute periods
0	10
1	35
2	19
3	25
4	6
5	2
6	3

The Planning Department's statistician claims that this distribution can be approximated by a Poisson distribution and recommends that no further surveys need be carried out at other filling stations. You are required to:

(a) discuss briefly the distinction between an empirical and a theoretical distribution in relation to the statistician's claim;

(b) test the validity of the statistician's claim showing all relevant workings.

The Chartered Institute of Public Finance and Accountancy.

5. The following data are available concening the products of a firm taken at random from four factories, and classified according to level of quality:

Levels of Quality	Factory A	Factory B	Factory C	Factory D
		(Numbers of items)		
Low	9	12	6	3
Average	22	52	30	16
Good	69	136	64	81

Is there any evidence to suggest that the level of quality is significantly associated with particular factories?

6. A study of the safety record of 50 international airlines over a selected period reveals the following incidence of aircraft "mishaps":

Number of mishaps	Number of airlines with this number of mishaps
0	13
1	15
2	11
3	7
4	3
5	1

Find the mean of this distribution, and calculate the frequencies given by the Poisson distribution with the same mean. How well does this new distribution fit the data?

(Take $e^{-1.5} = 0.224$).

The Chartered Institute of Transport.

CHAPTER 22

Analysis of Time Series

TIME SERIES

A time series has been defined as "data classified chronologically". It is a set of observations, for example, sales or production, over time (*see* Fig. 84.)

Amigo Cement Co. Ltd. production (Thousand tonnes)

	1st Qr.	2nd Qr.	3rd Qr.	4th Qr.
1976	83	81	98	114
1977	124	113	115	152
1978	163	162	168	175
1979	191	180	184	197

Fig. 84. An example of a time series.

CHARACTERISTICS OF A TIME SERIES

Two things are noticeable about the time series given in Fig. 84. First, there is a steady growth in production throughout the period. Secondly, there appear to be regular seasonal fluctuations, production declining during the second and third quarters of the firm's year.

A time series is the result of a number of movements. They are as follows.

(*a*) A *basic trend*. This will be the long-term movement. In Fig. 84, it is a steady growth.

(*b*) *Seasonal fluctuations*. These are deviations from the trend. They generally occur periodically, every quarter, week or month, according to the nature of the data.

(c) Catastrophic movements. These are caused by unusual events, for example, floods, strikes, fires, etc.

(d) Cyclical fluctuations. These are oscillatory movements super-imposed on the trend. In the case of economic and industrial time series they will correspond to movements of the trade cycle.

(e) Residual variations. These are the variations which are left after having removed all other movements and fluctuations. They will include chance variations arising from a multiplicity of causes.

The analysis of time series consists in separating these constituent parts.

THE USES OF TIME SERIES ANALYSIS

It enables forecasts to be made. If the trend line is continued into the future, as is shown by the broken line in Figs. 85, 86 and 87, a possible figure for some future date is obtained. Statisticians are not fortune-tellers, and the figure obtained by extrapolation, i.e. by projecting the trend line, is made on the assumption that the trend

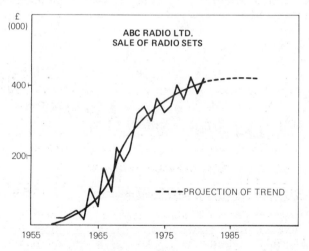

Fig. 85. A logistic trend.

will continue. Any unforeseen circumstances may invalidate the forecast. Other knowledge which may modify the projection must be taken into account. For the forecast to be of any value, the series from which it is obtained must be a long series. Despite the consider-able number of snags, forecasts based on the analysis of time series are useful in planning the various budgets in budgetary control.

If seasonal fluctuations are known, it is possible to adjust holdings of stocks so that unnecessary stocks are not held, or, on the other hand, to avoid having insufficient. It is an essential aid to planning, permitting perhaps other commodities to be produced or handled so that fluctuations are "evened out".

TRENDS

BASIC TRENDS

Three of the more usual basic trends are shown in Figs. 85, 86 and 87. The logistic trend has been found to be appropriate in many cases of natural growth, in the growth of production of certain commodities and in many types of business activity. The trend consists

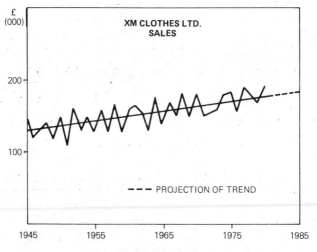

Fig. 86. An arithmetic trend.

of an increasing rate of growth, followed by a decreasing rate of growth, finally reaching a constant figure. In the case of sales, this would indicate that the market had reached saturation point.

Fig. 86 shows an arithmetic trend. This means that the variable increases or decreases by a constant amount each year.

Fig. 87 shows a compound interest trend. Such a trend will appear on semi-logarithmic paper as a straight line. This means that the variable will increase or decrease by the same proportion each year.

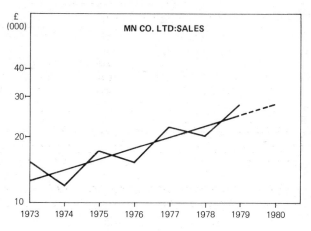

Fig. 87. A compound interest trend.

SEPARATION OF THE TREND

Among the methods possible are the following.

(a) Curve by inspection. Plot the series and draw a smooth curve by inspection in such a way that it lies between the points, so that the fluctuations in one direction are equal to those in the other direction, and show the general movement.

(b) The moving average method. This is the easiest method, but it has a number of disadvantages. It does not cover the complete period and usually it will not represent the trend quite faithfully.

(c) The method of least squares. A graph of the series will be required to enable a decision to be made as to the particular curve to be fitted.

THE CALCULATION OF MOVING AVERAGES

The number of items taken for each average will be the number required to eliminate the fluctuations; it will be the number required to cover the period over which the fluctuations occur. In Example 1, since the fluctuations are quarterly, a four-quarterly moving average is required. If monthly figures had been given, a twelve-monthly moving average would have been necessary. Where the period is not obvious, as in the case of cyclical fluctuations, which are not seasonal fluctuations, it will be necessary to draw a graph; the period taken is that between the "peaks" or, alternatively, between the "depressions" on the graphs.

Example 1.

Year	Production (tonnes)	5-Yearly total	Trend
1965	425	—	—
1966	526	—	—
1967	419	2,500	500
1968	487	2,750	550
1969	643	3,000	600
1970	675	3,250	650
1971	776	3,500	700
1972	669	3,750	750
1973	737	4,000	800
1974	893	4,250	850
1975	925	4,500	900
1976	1,026	4,750	950
1977	919	5,000	1,000
1978	987	—	—
1979	1,143	—	—

If a graph were drawn of the production through the period, it would be seen that "peaks" occur every five years; hence a five-yearly moving average is required. In this case, it is possible to see that "peaks" occur in 1966, 1971 and 1976, that is every five years, without actually drawing the graph. In actual practice, the trend would not, of course, be so regular.

The figures are calculated as follows:

Totals: $425 + 526 + 419 + 487 + 643 = 2,500,$
$526 + 419 + 487 + 643 + 675 = 2,750,$
$419 + 487 + 643 + 675 + 776 = 3,000,$

and so on.

Averages: $\dfrac{2,500}{5} = 500, \dfrac{2,750}{5} = 550,$ and so on.

The first total and moving average is opposite the third item, i.e. the middle item. There are no figures for the first two and the last two items. If there had been seven items to average, there would have been no figures for the first three, nor for the last three items, and the first total and average would have been opposite the fourth item. The case of an even number of items will be dealt with later.

FITTING A TREND LINE BY THE METHOD OF LEAST SQUARES

Only the case where the trend is appropriately represented by a straight line will be considered in this book. A graph is the best means of determining whether a straight line trend is appropriate.

Example 2.

Year	Sales (£000)	Time deviation from middle year	Deviations squared	Product of sales and deviations
1975	15	−2	4	−30
1976	17	−1	1	−17
1977	21	0	0	0
1978	23	1	1	23
1979	26	2	4	52
	102		10	28

Average sales $= \dfrac{102}{5} = 20.4$. This is the trend point value for 1977, the middle year.

The rate of growth $= \dfrac{\Sigma \text{ (Product of sales and deviations)}}{\Sigma \text{ (Deviations squared)}}$

$$= \frac{28}{10} = 2.8.$$

The trend is computed as follows:

```
1975  Average sales  (−2)(2.8) = 20.4 − 5.6 = 14.8.
1976        do.      (−1)(2.8) = 20.4 − 2.8 = 17.6.
1977            Average Sales           = 20.4.
1978  Average sales  (+1)(2.8) = 20.4 + 2.8 = 23.2.
1979        do.      (+2)(2.8) = 20.4 + 5.6 = 26.0.
```

A further example is given, this time with an even number of years. *Example 3.*

Year	Sales	Time deviation	Deviations squared	Product of sales and deviations
1974	12	−2.5	6.25	−30.0
1975	15	−1.5	2.25	−22.5
1976	17	−0.5	0.25	− 8.5
1977	21	+0.5	0.25	+10.5
1978	23	+1.5	2.25	+34.5
1979	26	+2.5	6.25	+65.0
	114		17.50	49.0

Average sales = 114/6 = 19. This is the trend point value for the middle of the period, i.e. half way between 1976 and 1977. Hence 1976 is half a year before the middle point and 1977 is half a year after the middle point.

Sales grow by an amount equal to 49/17.5 = 2.8.

The trend is as follows:

        1974 Average sales (−2.5)(2.8) = 19 −7.0 = 12.0.
        1975   do.   (−1.5)(2.8) = 19 −4.2 = 14.8.
        1976   do.   (−0.5)(2.8) = 19 −1.4 = 17.6.
        1977   do.   (+0.5)(2.8) = 19 +1.4 = 20.4.
        1978   do.   (+1.5)(2.8) = 19 +4.2 = 23.2.
        1979   do.   (+2.5)(2.8) = 19 +7.0 = 26.0.

When, instead of a constant amount of increase or decrease each year, there is a uniform percentage increase or decrease each year—that is to say, if the data were plotted on semi-logarithmic paper, a straight-line trend would be appropriate—the above method can still be used. In this case, however, logarithms are used as the basis of the calculation, instead of the original data. If it were required to fit such a compound-interest trend to the above data, logarithms of the sales would be substituted for the sales, the calculations would then be carried out in exactly the same way, and the trend would then be calculated in the form of logarithms. These would then be required to be transformed into numbers by means of tables. This method is useful for estimating increases of population as population increases according to the compound interest law.

SEASONAL FLUCTUATIONS

METHOD OF AVERAGES

This method ignores any trend that may be present. An example shows the necessary calculations.

Example 4.

Sales (£000)

Quarter	1977	1978	1979	Total	Percentage of each quarterly total to average quarterly total
1	3.6	3.7	4.2	11.5	104.5
2	2.5	3.4	3.0	8.9	80.9
3	3.9	3.7	3.1	10.7	97.3
4	4.6	4.5	3.8	12.9	117.3
				44.0	400.0

A much longer series is required for the estimates of seasonal fluctuations to be valid. Only three years were taken in the example to save long calculations.

The total in respect of each quarter is obtained. These are the figures given in the total column. The average quarterly total is then calculated. In Example 4, this is 44 divided by 4, which gives 11. The percentage of each quarterly total to average quarterly total is then worked out. Thus:

$$\text{Quarter 1} \left(\frac{11.5}{11}\right) 100 = 104.5;$$

$$\text{Quarter 2} \left(\frac{8.9}{11}\right) 100 = 80.9;$$

and so on.

If there are not the same number of years for each quarter, the average for each quarter will be calculated, and the average of these averages also calculated. The percentages of each quarterly average to the average quarterly average will then be computed. These will be the seasonal fluctuations.

Example 5.

Quarter	Total	Quarterly average	Percentage of each quarterly average to average quarterly average
1	11.5	$\dfrac{11.5}{3} = 3.8\dot{3}$	$\left(\dfrac{3.8\dot{3}}{3.6\dot{6}}\right)(100) = 104.5$
2	8.9	$\dfrac{8.9}{3} = 2.9\dot{6}$	$\left(\dfrac{2.9\dot{6}}{3.6\dot{6}}\right)(100) = 80.9$
3	10.7	$\dfrac{10.7}{3} = 3.5\dot{6}$	$\left(\dfrac{3.5\dot{6}}{3.6\dot{6}}\right)(100) = 97.3$
4	12.9	$\dfrac{12.9}{3} = 4.30$	$\left(\dfrac{4.30}{3.6\dot{6}}\right)(100) = 117.3$

Average quarterly average = $14.6\dot{6} \div 4 = 3.6\dot{6}$.

Had there not been the same number of years for each quarter, the totals would not all have been divided by the same number, e.g. 3. But, having obtained all the quarterly averages, the other calculations would have been precisely similar.

In the case of a monthly series, the procedure would be exactly the same; the total of the seasonal indices would, however, add up to 1,200 instead of 400.

METHOD OF MOVING AVERAGES

This method involves finding the trend, taking the trend from the original data, and averaging the resultant fluctuations.

Example 6. The data are from Fig. 84. A four-quarterly moving average is required to find the trend. This means averaging four items, so there will be no trend figure for the first two, nor for the last two items. Since a straightforward moving average would place the averages between the quarters, it is necessary to average adjacent pairs of the moving averages so that the trend values shall correspond to the same periods of the original data. This is done as follows:

Production (000 tonnes)	4-Quarterly total	Add in pairs	Trend (previous column divided by 8)
83	—	—	—
81	—	—	—
	376		
98		793	99
	417		
114		866	108
	449		
124		915	114
	466		
113		970	121
	504		
115	543	1,047	131
152	592	1,135	142
163	645	1,237	155
162	668	1,313	164
168	696	1,364	171
175	714	1,410	176
191	730	1,444	181
180	752	1,482	185
184	—	—	—
197	—	—	—

The four-quarterly figures are obtained as follows:

$$83 + 81 + 98 + 114 = 376,$$
$$81 + 98 + 114 + 124 = 417,$$
$$98 + 114 + 124 + 113 = 449,$$

and so on.

These totals, shown in the second column, are placed between the periods, since they relate to the mid-point.

The "add in pairs" figures are obtained as follows:

$$376 + 417 = 793,$$
$$417 + 449 = 866,$$
$$449 + 466 = 915,$$

and so on.

The first total is placed opposite the third quarter, the next total opposite the next quarter, and so on.

The "sums in pairs" are divided by 8. The resultant figures give the trend. They are shown in column 4.

In the case of monthly figures, a twelve-monthly moving total would be required. They would then be summed in pairs and the sums divided by 24. The first six items and the last six would have no corresponding trend value.

The fluctuations from the trend are now averaged to give the seasonal fluctuations. These fluctuations can be regarded as the difference between the trend and the original data or, alternatively, production as a percentage of the trend. This latter view is more reasonable if the trend is changing rapidly. The table below shows the necessary calculations.

Production

Production	Trend	Fluctuations (production less trend)	Percentage production to trend
1976 3rd Qr. 98	99	−1	99.0
4th Qr. 114	108	+6	105.5
1977 1st Qr. 124	114	+10	108.8
2nd Qr. 113	121	−8	93.4
3rd Qr. 115	131	−16	87.8
4th Qr. 152	142	+10	107.0
1978 1st Qr. 163	155	+8	105.1
2nd Qr. 162	164	−2	98.8
3rd Qr. 168	171	−3	98.2
4th Qr. 175	176	−1	99.4
1979 1st Qr. 191	181	+10	105.5
2nd Qr. 180	185	−5	97.3

The next stage in obtaining the seasonal fluctuations is to average the fluctuations as computed in the manner just shown. These fluctuations include random fluctuations as well as the seasonal fluctuations. Averaging has the effect of eliminating random fluctuations.

If the seasonal fluctuations are considered as the difference between the original data and the trend, they will be computed as follows:

	1st Qr.	2nd Qr.	3rd Qr.	4th Qr.
1976	—	—	− 1	+ 6
1977	+10	− 8	−16	+10
1978	+ 8	− 2	− 3	− 1
1979	+10	− 5	—	—
	+28	−15	−20	+15

	1st Qr.	2nd Qr.	3rd Qr.	4th Qr.
Unadjusted average	+ 9.3	− 5.0	− 6.7	+ 5 = + 2.6
Adjustment	− 0.65	− 0.65	− 0.65	− 0.65 = − 2.6
Seasonal fluctuations	+ 8.65	− 5.65	− 7.35	+4.35 = 0.0

The fluctuations for each quarter are averaged. These fluctuations should add up to zero. As will be noted, in the example given they add up to 2.6. The adjustment consists in subtracting one quarter of this difference from each quarter, thus making the adjusted averages add up to zero. These adjusted averages are the seasonal fluctuations.

If the seasonal fluctuations are considered as a percentage of the trend, the seasonal indices, as they are then called, are computed as follows:

	1st Qr.	2nd Qr.	3rd Qr.	4th Qr.
1976	—	—	99.0	105.5
1977	108.8	93.4	87.8	107.0
1978	105.1	98.8	98.2	99.4
1979	105.5	97.3	—	—
	319.4	289.5	285.0	311.9

	1st Qr.	2nd Qr.	3rd Qr.	4th Qr.
Unadjusted average	106.5	96.5	95.0	104.0 = 402.0
Seasonal index	105.9	96.0	94.6	103.5 = 400.0

The fluctuations for each quarter are averaged. These should add up to 400. As will be noted, in the table above they add up to 402. The adjusted averages are obtained by multiplying each average by

400 and dividing by 402, thus making the adjusted averages add up to 400.

In the case of monthly seasonal indices they will add up to 1,200.

Seasonal indices are a more logical way of dealing with seasonal fluctuations.

TESTING THE SUITABILITY OF A SEASONAL INDEX

A separate graph is required for each quarter or month, as the case may be. The years are plotted along the "horizontal" axis. The percentages of the moving averages or the fluctuations (original data less trend) are plotted along the "vertical" axis. Any trends show that the method of computation of the seasonal indices or seasonal fluctuations is not suitable; changing rather than stable indices are required.

PROCEDURE

Computing seasonal indices.

1. Calculate the moving average, the number of items being such as to eliminate the fluctuations, to obtain the trend.
2. Calculate the original data as a percentage of the trend.
3. Tabulate the results of step 2: a column for each period of fluctuation, e.g. month, quarter, and a row for each period which eliminates the fluctuations, e.g. 1980, 1981.
4. Average the values of each column.
5. Adjust the figures of step 4 so that they add up to a "correct" total, e.g. 1,200 in the case of monthly fluctuations. These are the seasonal indices.

DESEASONALISED DATA

If the seasonal fluctuations are deducted from the original data, the result is the deseasonalised data. This will consist of the trend and the residual fluctuations. Alternatively, the original data is divided by the seasonal indices to give the deseasonalised data. An example of the first method, using the same data as before, is shown below.

Production

		Pro- duction	Seasonal fluctua- tions	Deseason- alised data	Trend	Residuals
1976	3rd Qr.	98	−7.4	105.4	99	6.4
	4th Qr.	114	+4.4	109.6	108	1.6
1977	1st Qr.	124	+8.6	115.4	114	1.4
	2nd Qr.	113	−5.6	118.6	121	−2.4
	3rd Qr.	115	−7.4	122.4	131	−8.6

FORECASTING

SIMPLE FORECASTING

The figures of production of Amigo Cement Co. Ltd. are given up to and including the fourth quarter of 1979. It is required to forecast the production for the first two quarters of 1980. The procedure is as follows.

(a) Find the trend value for the first two quarters of 1980.

(b) Add the appropriate seasonal variations or, better still, multiply by the appropriate seasonal indices.

A simple way of finding the trend values is shown below.

	Trend		Trend
3rd Qr. 1977	131	3rd Qr. 1978	171
4th Qr. 1977	142	4th Qr. 1978	176
1st Qr. 1978	155	1st Qr. 1979	181
2nd Qr. 1978	164	2nd Qr. 1979	185
	592		713

Mean value of trend $\frac{592}{4} = 148$ $\qquad$ $\frac{713}{4} = 178$

Increase in mean value over one year $= 178 - 148 = 30$

Increase in mean value per quarter $= \frac{30}{4} = 7.5$

Trend

2nd Qr. 1979		185
3rd Qr. 1979 = 185 + 7.5	= 192.5
4th Qr. 1979 = 192.5 + 7.5	= 200
1st Qr. 1980 = 200 + 7.5	= 207.5
2nd Qr. 1980 = 207.5 + 7.5	= 215

The forecasts for the first two quarters of 1980 are:

1st Qr. 1980; $(207.5)\left(\dfrac{105.9}{100}\right)$ = 220 thousand tonnes.

2nd Qr. 1980; $(215)\left(\dfrac{96}{100}\right)$ = 206 thousand tonnes.

EXPONENTIAL SMOOTHING

This is a method of short-term forecasting. The only information needed is *(a)* the forecast for the current period—the old forecast—and *(b)* the value for the current period—the observation. It is then possible to make a forecast for the next period—the new forecast.

The new forecast is calculated as follows:

NEW FORECAST = OLD FORECAST + α (OBSERVATION − OLD FORECAST)
where α is the *smoothing factor* or *smoothing constant.*

If $\alpha = 0$, then the new forecast = the old forecast, that is, there is complete stability; if $\alpha = 1$, then the new forecast = latest observation, that is, there is complete sensitivity. In most cases a smoothing factor between 0.1 and 0.2 will be found suitable. The smaller the value of the smoothing factor, the smoother the forecast series, that is, the smaller the deviations.

The monthly production of a firm is as follows: January 38,000 tonnes, February 34,000, March 40,000, April 36,000, May 32,000, and June 38,000. The table below shows the calculations of the forecasts, and Fig. 88 enables a comparison to be made between the actual production and the forecasts.

In the example just given there is no trend. Where a trend exists a more complicated formula is required. To the formula for exponential smoothing of the observations are added terms to provide for the exponential smoothing of the trend.

Monthly Production (thousand tonnes)

	Old forecast (a)	Observation (b)	Error (observation— old forecast) (c) (b) − (a)	α (error) (α = 0.2) (d) (α) × (c)	New forecast (e) (d) + (a)
Jan.	32.0	38	6.0	1.2	33.2
Feb.	33.2	34	0.8	0.2	33.4
Mar.	33.4	40	6.6	1.3	34.7
Apr.	34.7	36	1.3	0.3	35.0
May	35.0	32	−3.0	−0.6	34.4
June	34.4	38	3.6	0.7	35.1

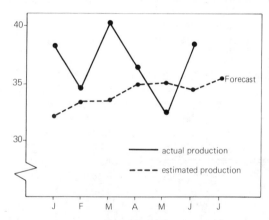

Fig. 88. Exponential smoothing of a time series.

PROCEDURE

Short-term forecasting

1. Choose a value for α, the smoothing constant.
2. Deduct the *forecast* for the *current* period—that is the "old" forecast—from the actual *observation* for the *current* period.
3. Multiply the error obtained in step 2 by α.
4. Add the product obtained in step 3 to the "old" forecast to obtain the new forecast, i.e. the *forecast* for the *next* period.

WHEN THE DATA REQUIRE ADJUSTMENT BEFORE ANALYSIS

Over the past thirty years the value of money has continually fallen. Any time series expressed in terms of money will show this change in addition to the actual changes in the data. In most cases the data will require to be deflated by a suitable price index, often specially computed for the purpose, before analysing the time series, in order to eliminate this effect of the changing value of money.

In the case of monthly figures, adjustments may be necessary because of the varying length of the month. The original data is multiplied by an appropriate correction factor.

Examples of such factors are:

(a) $\dfrac{\text{average number of days per month}}{\text{actual number of days in the month}}$

(b) $\dfrac{\text{average number of working days per month}}{\text{actual number of working days in the month}}$

QUESTIONS

1. From the data below compute the trend, stating your method of determining the period of the cyclical fluctuations.

Year		Year		Year		Year	
1957	11.1	1963	11.4	1969	10.0	1975	12.6
1958	13.1	1964	11.5	1970	10.3	1976	10.1
1959	8.3	1965	11.7	1971	13.0	1977	12.4
1960	10.0	1966	12.3	1972	11.4	1978	10.7
1961	10.1	1967	11.0	1973	12.3	1979	12.9
1962	9.2	1968	9.4	1974	14.2		

2. The following figures are the quarterly estimates of a country's Personal Expenditure on Fuel and Light (£000,000):

	I	II	III	IV
1972	78	62	56	71
1973	84	64	61	82
1974	92	70	63	85
1975	100	81	72	96
1976	107	81		

Calculate the average seasonal movement in this series allowing for any trend which may be present.

Chartered Institute of Transport.

3.

Personal Expenditure on Fuel and Light
£(million)

Quarters	1	2	3	4
1977	105	77	68	95
1978	107	83	74	106
1979	117	99	86	112

Explain how seasonal variations are removed from time series. Graph the series above and insert the trend of expenditure.

4. Write brief notes on:

      (a) class interval   (b) compensating error
      (c) primary date   (d) smoothing of curves

Institute of Cost and Management Accountants.

5. The following table shows the amount of money (£000,000) spent upon passenger travel in Barbalonia at levels of fares, and charges current, during the periods given:

	1975	1976	1977	1978	1979	1980
Quarter I	71	71	73	76	84	89
II	89	90	91	97	106	110
III	106	108	111	122	130	—
IV	78	79	81	89	95	—

Calculate the seasonal pattern for this series and estimate the values of the series for the third and fourth quarters of 1980.

6. The following table shows the amount of money (£000,000) spent upon passenger travel in Barbalonia for the periods given when valued at the level of fares and charges in force during 1975.

	1975	1976	1977	1978	1979	1980
Quarter 1	71	71	73	71	74	76
II	89	88	91	89	91	94
III	106	104	110	113	112	—
IV	78	78	80	81	81	—

Using this series and the figures in Question 5, plot a scatter chart.

What general information does this afford on the question of changes in the levels of fares and charges?

7.

Saleable Output of Open-cast Coal
(in thousands of tonnes per quarter)

Year	I	II	III	IV
1975	–	–	–	2,595
1976	2,340	3,051	3,061	2,565
1977	2,458	3,161	3,466	2,793
1978	2,809	2,911	3,035	2,917
1979	2,652	2,776	2,634	–

From the above table:

(a) compute the mean seasonal variation present;

(b) rewrite the figures for the year 1979 with the seasonal variation removed.

8.

Coal-Mining Industry
Number of Newly Employed Juveniles under Eighteen Years

	Quarters			
	1	2	3	4
1964	3,934	3,792	3,860	2,569
1965	3,872	2,823	4,522	4,902
1966	7,178	3,503	5,411	3,493

Rewrite these numbers to the nearest hundred and then calculate the trend and seasonal variation.

9. Assuming the true trend of the number of trucks and wagons owned by the main line railway companies was represented by the years 1965 to 1974—estimate the figure for 1980.

End 1965	634	End 1973	683
1966	634	1974	687
1967	657	1975	678
1968	664	1976	665
1969	664	1977	656
1970	670	1978	689
1971	672	1979	691
1972	678	1980	(?)

10.

Total Production of Paper
Weekly averages (thousand tonnes)

Year	Quarters			
	1	2	3	4
1966	37	38	37	40
1967	41	34	25	31
1968	35	37	35	41

Calculate the moving average trend of the above data and estimate the seasonal variation in the weekly paper production.

Draw on the same diagram the weekly average production together with the trend.

11. The production of small cigars in the United States 1961—70 is shown in the following table.

Year	Number of small cigars (millions)
1961	98.2
1962	92.3
1963	80.0
1964	89.1
1965	83.5
1966	68.9
1967	69.2
1968	67.1
1969	58.3
1970	61.2

You are required to:

(a) graph the data shown above,

(b) calculate the equation of a least squares trend line fitting the data, say what the trend value is at 1961 and 1970, and explain your result, and

(c) estimate the production of small cigars for the year 1971.

Association of Certified Accountants.

12. Smooth exponentially the following time series and forecast the profits for August 1976. Use a smoothing constant of 0.1.

The profits of Ansco were: January 1976 £52,000, February £64,000, March £56,000, April £54,000, June £60,000, July £54,000.

(*Note.* At the start of a series any reasonable value may be taken as the "old forecast". Let it be £52,000 in this case.)

13. Quarterly production for a paper-making plant was reported as follows:

Years	Quarterly production ('00 tonnes)			
	I	II	III	IV
1972	38.8	41.3	39.0	45.6
1973	44.7	45.2	42.0	49.9
1974	46.7	48.2	44.5	51.3
1975	50.1	54.6	—	—

Using the method of moving averages, find the average seasonal deviations and thus estimate the production figures for the last two quarters of 1975.

> *Institute of Chartered Secretaries and Administrators.*

14. Required:

In the analysis of time series:

(*a*) explain what is meant by a trend line and describe very briefly two principal methods of calculating it;

(*b*) explain what is meant by residual variations and why they are important;

(*c*) give two situations where it is important to adjust the data prior to analysis, with reasons;

(*d*) sketch a graph of a time series which shows a trend and also cyclical and seasonal movements.

> *Association of Certified Accountants.*

15. New rentals of colour television sets were reported by one of the suppliers for the year 1977 to the middle of 1980 as follows:

Number of New Rentals (000s)

Years	Quarters			
	I	II	III	IV
1977	26.2	30.4	29.6	32.7
1978	31.3	36.8	30.5	37.9
1979	35.0	40.2	29.6	38.4
1980	30.6	35.5		

Using the method of moving averages, find the average seasonal deviations, and thus estimate the new rentals for the last two quarters of 1980.

16. What is the significance of the smoothing constant (α) in the technique of exponential smoothing? Using an α value of your choice, apply the technique to the following observations, if an initial forecast of 55 is given:

51, 56, 53, 52, 54, 56, 49.

Chartered Institute of Transport.

CHAPTER 23

Statistical Quality Control

Statistical quality control is a method of estimating the quality of the whole from the quality of the samples taken from the whole. The method is based upon the laws of chance and has a sound mathematical basis.

Among the many advantages of statistical quality control, the most important is that it is more efficient. 100 per cent inspection (i.e. inspecting every article) is not reliable; the inspector usually inspects as many as possible and passes the rest. Human fatigue also causes errors of inspection. Quality control charts, on the other hand, indicate not only whether the quality is actually up to the required standard, but, as will be seen later on in the chapter, give warning of possible future lack of quality which can hence be avoided. Another important advantage may be a saving in cost. Since only a part of the whole is inspected, fewer inspectors are needed; there is a saving in time and wages.

THEORETICAL FOUNDATION OF QUALITY CONTROL

A machine may be set to cut out a hole of a given diameter in, say, a piece of metal. Yet although there is no cause, such as a fault developing in the machine or the operator becoming tired or careless, the measurements will vary. For this reason tolerance limits are fixed. There are the largest and smallest sizes between which the measurement must fall to satisfy the required specification. When there is no assignable cause, i.e. a specific reason for variations in measurements such as those just mentioned, the measurements of the pieces will vary according to the "normal" distribution.

This "normal" distribution has the following characteristics.

(a) 998 out of every 1,000 pieces will vary in size between the average size plus 3.09 times the standard deviation and the average size less 3.09 times the standard deviation. This is expressed in symbols as $\overline{X} \pm 3.09\ \sigma$ (pronounced X bar plus or minus 3.09 sigma), $\overline{X}$ standing for the average size and σ standing for the standard deviation.

(b) 38 out of every 40 pieces will vary between the average size plus 1.96 times the standard deviation and the average size less 1.96 times the standard deviation.

This distribution occurs when pure chance alone causes variations in the sizes. If, therefore, more than one piece in a thousand were larger than $\overline{X} + 3.9\ \sigma$, it would be necessary to find an assignable cause, one that can be removed. Variations due to pure chance cannot be removed.

The entirety of the batch is known as the population. The average size of the population is denoted by $\overline{X}$ and the standard deviation by σ_p (the "$_p$" denoting population).

However, the only information available will be from the samples taken. No information can be obtained from a single sample, although in practice a population is often judged by a single sample, which is most unscientific and completely unreliable. The average measurement of the samples is considered the best estimate of the average measurement of the population. The greater the number of samples the better the estimate.

The measurement of each piece in each sample is taken. In respect of each sample it will therefore be possible to find the range of measurements of each sample and also the average measurement of each sample. This average measurement of each sample will also vary according to the "normal" distribution (again provided there is no assignable cause why it should not). The averages have, however, as their standard deviation $\sigma_p/\sqrt{n}$, where n is the number of items in each sample—the number of items in each sample will always be the same.

The variation in the sizes of the individual items and the variation in the average measurements of the samples *must not be confused*.

The average measurements of the samples, since they vary according to the "normal" distribution, will have as their limits (apart from 2 in every 1,000) $\overline{X} \pm 3.09\ \sigma_p/\sqrt{n}$. These are known as the control limits for means.

$3.09\ \sigma_p/\sqrt{n}$ is calculated by multiplying the average range of the samples, denoted by $\overline{w}$, by an A factor. This factor varies according to the number in the sample. If the sample averages fall outside the limits, the individual items' measurements will also fall outside their limits. It will also be necessary to look for an assignable cause.

The ranges of all the samples might vary considerably, and yet the average range still be within the permissible limit. The calculation of the variation of the sample measurements is based on the average range. It is therefore necessary to plot the ranges of the individual samples on a range chart and verify that they do not fall outside the control limit of ranges. This control limit is arrived at by multiplying the average range by a *D* factor. This factor also varies according to the number of items in the sample. In the case of ranges, only the upper limit is of relevance. In actual practice it has been found unnecessary to continue plotting the range chart when it has been found satisfactory for a comparatively short period.

CONTROL CHARTS

The means control chart enables possible trouble to be averted. The principal indications when action should be taken are:

(*a*) an average outside the control limits,
(*b*) several averages, especially if consecutive, near a control limit,
(*c*) an undue number of averages above or below a central value,
(*d*) a trend in averages,
(*e*) an unusually long run of averages above or below a central value.

The first step in the making of control charts is to collect the data and enter the data on sheets. Measurements are taken in respect of each component in each sample. These are entered on sheets, together with details of tolerance limits, date and time of samples, operator's number, the inspector's name, details of the component and the dimension being controlled, and any other relevant details. Samples are taken at suitable convenient periods.

The measurements in respect of each sample will be added and divided by the number in the sample. This will give the average measurement of the sample. The largest measurement in each sample less the smallest measurement will give the sample range. These figures will also be entered on the data sheet.

The next step is to plot these sample means and ranges on two separate graphs both drawn on the same sheet of paper. The scales should be carefully chosen to suit the particular measurements of the component's dimensions being dealt with. The scale is shown on the "vertical" axis. The tolerance limits should also be marked on the means chart.

The third stage is to insert temporary control limits. It has been suggested (*A First Guide to Quality Control for Engineers,* H.M.S.O.) that temporary control limits be fixed after ten samples have been taken. This is done as follows.

(a) The ten sample means are averaged to find the grand mean, denoted by $\overline{X}$.

(b) The ten sample ranges are averaged to give average range, denoted by $\overline{w}$.

(c) $A\overline{w}$ and $D\overline{w}$ are computed. The values of A and D will be found in Appendix F.

(d) Draw horizontal lines on the means chart at (i) $\overline{X}$, (ii) $\overline{X} + A\overline{w}$, (iii) $X - A\overline{w}$. Lines (ii) and (iii) are the upper and lower control limits respectively.

(e) Draw horizontal lines at $\overline{w}$ and $D\overline{w}$. The latter line is the control limit for ranges. Since these control limits are temporary, they should be drawn in pencil.

The final stage is to compute new control limits after a further 15 samples have been taken, making 25 in all. The whole of the data of the 25 samples should be used. These final limits should be entered in red ink, the averages being in blue.

Once control limits have been established, no point should be allowed to fall outside without investigating the reason. When points no longer fall outside, the process is said to be under control.

STATISTICAL CONTROL AND TOLERANCE LIMITS

A machine may be turning out parts whose variations in size are entirely due to pure chance. The process is under statistical control. Nevertheless the variations may be greater than those necessary to comply with the specification. *It is therefore important to make certain that the variation in the sizes of the individual items is not greater than the variations allowed by the tolerance limits.*

The tolerance range—the difference between the largest size and the smallest allowed by the specification—is multiplied by an L factor, which varies with the number of items in the sample. Provided the average range of the samples does not exceed this figure, the variation of the individual items coming from the machine is within the variation allowed by the specification. If, however, $\overline{w}$ exceeds L times the tolerance range, the control chart will be unsatisfactory; $\overline{w}$ *must not exceed* $L(T_U - T_L)$. The alternatives are 100 per cent inspection or the purchase of a more precise machine.

One further test remains before the control chart can be considered satisfactory. The amount of variation may be satisfactory, but it must also be between the specified measurements. To test this it is necessary to verify that the control limits lie between $T_U - M\overline{w}$ and $T_L + M\overline{w}$, where T_U is the upper tolerance limit, and T_L is the lower tolerance limit and M is a factor which varies according to the

number of items in the sample. If the control limits do not lie within this band, but the tolerance range is satisfactory, the machine
will require resetting.

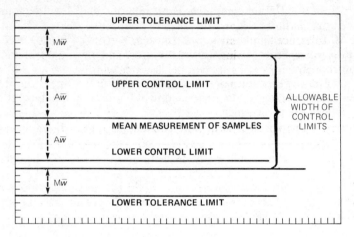

Fig. 89. *The relationship between control limits, tolerance limits and allowable width of
control limits.*

If the control limits are within the allowable width, but are widely
separated from them, the machine being used for the process is too
precise. A cheaper and less precise machine might be used.

MEANS CHARTS WITH TWO SETS OF CONTROL LIMITS

The 3.09 sigma limits or 1/1,000 limits are the more usual. Sometimes, however, 1/40 limits are used in addition. When a reading falls
outside these latter limits it is taken as a warning, and an investigation takes place. When, however, a reading falls outside the former,
it is considered that there is definitely something wrong.

Example 1. The internal diameter of a ring is required to be
3 mm, with a tolerance of 0.24 mm. Samples of 4 are taken every
half-hour. The first samples were found to have a mean measurement of 3.01 mm and an average range of 0.12 mm. The values of
the factors *A, L, M* and *D* for samples of 4 are 0.75, 0.33, 0.75 and
2.57 respectively. Draw a means chart and a ranges chart and verify
if they are satisfactory.

(a) Calculation of control limits for means.
Upper control limit: 3.01 + (0.75)(0.12) = 3.10 mm.
Lower control limit: 3.01 − (0.75)(0.12) = 2.92 mm.

(b) Verification that variation of sizes of items is not greater than tolerance spread.

Difference between greatest and smallest size allowable: 0.48 mm.

L (0.48) = (0.33)(0.48) = 0.16. Since 0.12 does not exceed 0.16 the variation of individual sizes is satisfactory.

(c) Calculation of allowable width of control limits.

Upper tolerance limit less $M\bar{w}$ = 3.24 − (0.75)(0.12) = 3.15.

Lower tolerance limit plus $M\bar{w}$ = 2.76 + (0.75)(0.12) = 2.85.

Control limits for means are within allowable widths.

(d) Calculation of control limit for ranges.

Control limit: $D\bar{w}$ = (2.57)(0.12) = 0.30.

On the charts shown in Figs. 90 and 91 measurements are plotted relating to the period subsequent to the taking of the 25 samples.

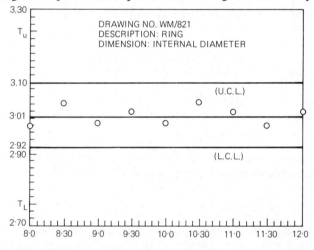

Fig. 90. A means chart.

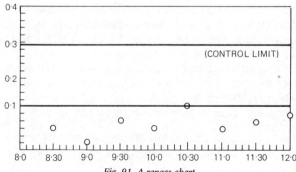

Fig. 91. A ranges chart.

PROCEDURE

Compiling means and ranges charts.

1. Decide the number in the sample.
2. Decide how often a sample should be taken.
3. Record the measurements in respect of each item in each sample.
4. In respect of each sample calculate *(a)* the mean, and *(b)* the range.
5. Calculate the grand mean (denoted by $\overline{X}$) and the average range denoted by $\overline{w}$).
6. Calculate the U.C.L. and the L.C.L., i.e. $\overline{X} \pm A\overline{w}$.
7. Verify that variations in measurements do not exceed tolerance spread, i.e. that $\overline{w}$ does not exceed $L(T_U - T_L)$.
8. Calculate allowable width of control limits, i.e. $T_U - M\overline{w}$ and $T_L + M\overline{w}$.
9. Verify that control limits are within allowable width.
10. Draw horizontal lines on the means chart at $\overline{X}$, $\overline{X} + A\overline{w}$ and $\overline{X} - A\overline{w}$.
11. Calculate control limit for ranges, i.e. $D\overline{w}$.
12. Draw horizontal lines on ranges chart at $\overline{w}$ and $D\overline{w}$.

FRACTION DEFECTIVE CHARTS

Much quality inspection takes the form of determining whether a product or component is acceptable or defective. If the proportion of defective items is constant over a production run, then the proportions of defectives (now named *fraction defective*) in samples of the same size will be distributed binomially.

The *fraction defective control chart* will have for its *upper control limit* the estimated proportion of defectives in the production run *plus* three standard errors of a proportion and for its *lower limit* the estimated proportion of defectives *less* three standard errors *or* zero, whichever is the greater. Any proportions falling outside these limits are assumed not to be the result of a sampling variation but due to an assignable cause.

It is important that large samples are used in the construction and use of fraction defective control charts; small samples give unsatisfactory results.

Example 2. Twenty five samples of 100 each are taken of a certain component of a diesel engine. The number of defectives were

5, 3, 6, 2, 7, 2, 6, 2, 4, 6, 3, 8, 4, 7, 8, 2, 0, 4, 5, 4, 2, 2, 7, 3, 8.

Make a fraction defective control chart and plot the fraction defective in respect of the number of defectives found in samples taken subsequently to the twenty five samples required to make the chart. The number of defectives found in the samples were

$$5, 3, 6, 4, 5, 7.$$

Total number of defectives = 5 + 3 + 6 + . . . 3 + 8 =   110
Total number inspected = 25 × 100   = 2,500
Estimated proportion in production run = $\dfrac{110}{2,500}$ = 0.044

$p = 0.044$; $q = 1 - 0.044 = 0.956$; $n = 100$.

$$\text{s.e.}_{\text{prop}} = \sqrt{\frac{(0.044)(0.956)}{100}} = 0.02$$

U.C.L.   = 0.044 + 3(0.02)   = 0.104
L.C.L.   = 0

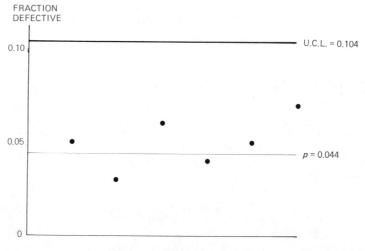

Fig. 92. Fraction defective chart.

Figure 92 shows the fraction defective control chart. It shows that the results of subsequent samples plotted do not fall outside the limits but there does seem to be a tendency as shown by the last three results plotted for the fraction defective to increase.

PROCEDURE

Compiling a fraction defective chart.

1. From a large number of samples compute the fraction defective.
2. Compute the standard error of the proportion.
3. Determine the control limits.
4. Draw horizontal lines on chart at control limits.

CUSUM CHARTS

The means chart (Fig. 90) indicates when corrective action should be taken but a cumulative sum (cusum for short) chart will show a change in the mean value much more quickly. The change appears on the chart as a change in the slope of the cusum graph.

PROCEDURE

Compiling a cusum chart.

1. Choose a target value denoted by T, e.g. the specification value.
2. Compute the deviations of the individual values x from the target value, i.e. $x - T$.
3. Compute $\Sigma(x - T)$, i.e. the cusum.
4. Plot the cusum (y-axis) against each successive item (x-axis).

The deviations of the individual values from the target value will fluctuate around the target value, some negative and some positive, and the cumulative sum of these deviations will approximate to zero, provided that the process is under control and the target value is near the mean measurement of the product coming from the machine. When there is a change in the slope of the chart, a change in the mean value is indicated.

Cusum charts are not only useful as a technique in quality control, but are useful in indicating changes in, amongst others, the level of sales, production, stocks or index numbers.

Example 3. The last hundred components off the cutting machine had a mean length of 613 mm. This was satisfactory but it was important that the mean length should not change. The lengths of the next 16 components coming off the machine were:

615, 614, 613, 613, 610, 615, 618, 611, 610, 611,
617, 616, 612, 621, 612, 618 mm.

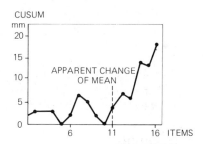

Fig. 93. Cusum chart for data of Example 3.

Target (T) = 613 mm			
Item number	x mm	x−T mm	Cusum mm
1	615	2	2
2	614	1	3
3	613	0	3
4	613	0	3
5	610	−3	0
6	615	2	2
7	618	5	7
8	611	−2	5
9	610	−3	2
10	611	−2	0
11	617	4	4
12	616	3	7
13	612	−1	6
14	621	8	14
15	612	−1	13
16	618	5	18

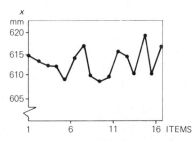

Fig. 94. Coventional chart for data of Example 3.

Draw a cusum chart and indicate where there is any change in the mean. Also draw a conventional chart and compare the two.

Figure 93 shows the cusum chart for the data given. It would appear that from item 11 onwards the mean length of the components has increased. A conventional chart is shown in Figure 94. The change in the mean value is not apparent.

QUESTIONS

For factors refer to Appendix F.

1. What do you know of "Statistical Quality Control"? Give some account of its basis and methods.

2. A piece of metal is required 10.75 cm long. The tolerance allowable is 0.15 cm either way. The average length of a large number of samples was 10.76 cm. The average range was 0.08 cm. Each sample consisted of 5 pieces. You are required to draw a means control chart and to verify if it is satisfactory.

3. What advantages does statistical quality control possess over 100 per cent inspection of production or sampling which is not based on statistical theory?

4. At a factory pink and white sweets are produced in separate plants. Approximately equal streams of each colour are fed into a single large mixer, where the sweets are well mixed. Batches, each of 100 sweets, are drawn from the mixer and a count is made of the number of sweets of each colour. 400 batches were examined and the following information was obtained:

Average number of pink sweets, per batch   50
Standard deviation of the number of pink sweets per batch   5

(a) What is the standard deviation of the number of white sweets per batch?

(b) What is the maximum number of pink sweets that you would expect to find in any one batch taken at random?

(c) If the sweets were drawn off in batches of 400 what would you expect the average number of pink sweets per batch (of 400) to be, and what would you expect the standard deviation of the number of pink sweets per batch to be?

(d) For batches of 100, draw a control chart on which the results of the colour counts could be entered. Enter on this chart points which would indicate that the mixing was at first satisfactory, but had then become unsatisfactory. Give your reasons in each case.

Institute of Cost and Management Accountants.

5. The interior diameter of a ring is required to be 120 mm with an allowable tolerance of 0.25 mm either way. The average diameter of a large number of samples was 119.97 mm and the average range was 0.13 mm. Each sample consisted of 6 rings.

You are required to:

(a) draw a means control chart showing both action and warning limits and say whether the chart is satisfactory.

(b) draw a ranges chart.

CHAPTER 24

Statistics and Business Management

STATISTICS IN BUDGETARY CONTROL

Budgets are predetermined plans. The estimates of sales will be the basis of the sales budget, estimates of production, the basis of the production budget, estimates of expenses, the basis of the expense budget, and so on. The budgets will be analysed according to departments, products, areas and any other relevant analysis. Periodic reports will be made, and actual results compared with budgeted plans, which will be continuously revised as further information becomes available and conditions change. The budgets are the basis of control of the activities of a firm.

Generally, the sales budget is the basis of all the other budgets. For this reason the estimate of future sales is most important. One method is the time series analysis dealt with in Chapter 22. The trend is plotted on graph paper; a smooth curve is drawn between the points (this may, of course, be a straight line); this trend is continued until it reaches the date for which the forecast is being made; the extrapolated trend is multiplied by the appropriate seasonal index; this figure will be the required estimate, subject to any modification called for by any changing conditions. Before the sales figure is adopted for the sales budget, it must be considered from a number of points of view. Perhaps additional plant might be necessary. Would it be possible to finance the budget?

MANAGEMENT RATIOS

The ratios given below are compiled at suitable intervals. If they are charted on graphs, movements in their value can be noted and the appropriate action taken. It might not be out of place to mention that graphs are used to enable action to be taken, or to refrain from taking action, as the case may be. They are not to be used as pretty pictures.

(a) The 100 per cent statement (Balance Sheet). Each asset is shown as a percentage of the total assets; each liability is shown as a percentage of the total liabilities. Balance Sheets of different undertakings or of the same undertaking at different dates can then be compared.

(b) The 100 per cent statement (P. and L. a/c). The costs shown in the Profit and Loss Account are shown as percentages of net sales. The appropriation of net profit to its various uses can also be given on a percentage basis.

(c) Ratio of current assets to current liabilities. This is an extremely important ratio. Its purpose is to enable a firm to know if it is able to meet its current liabilities. The ratio depends on the rate at which stock is converted into cash. The ratio will vary according to the season.

(d) Stock turnover. This ratio will measure the rate at which stock is converted into cash or debtors. The formula is:

$$\frac{\text{Cost of goods sold in a given period}}{\text{Average stock of goods during period}}$$

Alternatively, the turnover ratio may be expressed as the number of days required to sell the stock. This is computed by dividing the number of days in the given period by the ratio as given above. The stock turnover should be computed for every type of commodity. There will be seasonal variations. This ratio measures the efficiency of the buying department.

(e) Debtors' turnover. This ratio will measure the rate at which debtors are turned into cash. The formula is:

$$\frac{\text{Net credit sales in a given period}}{\text{Average value of debtors during period}}$$

Alternatively, if the number of days in the given period are divided by the above ratio, the turnover is expressed as the number of days required to collect debts.

Comparison of this figure with the conditions of sale enables the efficiency of the collecting department to be judged.

(f) Ratio of net worth to fixed assets. A fall in this ratio indicates an expansion of fixed assets, probably financed by borrowing.

(g) Ratio of sales to fixed assets. This ratio will enable a firm to know if an increase in fixed plant is producing correspondingly greater sales.

(h) Ratio of net worth to liabilities. The ratio of net worth, i.e. shareholders' capital plus accumulated undistributed profits to liabilities indicates to what extent creditors are financing the business. If the ratio were 4, this would indicate four-fifths of the capital was provided by the shareholders and one-fifth by the creditors.

(i) Ratio of net profit to net worth. This ratio indicates whether the funds invested in the business are earning a reasonable return.

(j) Labour turnover. The cost of replacing workers who leave is high. Labour turnover must therefore be kept at a minimum. A high labour turnover may indicate bad management, discontent or inefficient recruitment. The labour turnover ratio must therefore be continually reviewed. The two principal formulae are:

$$\frac{\text{Number of workers replaced during given period}}{\text{Average number of workers during period}}$$

or

$$\frac{\text{Number of workers who leave in given period}}{\text{Average number of workers during period}}$$

The second formula enables the ratio to be analysed into:

$$\frac{\text{Resignations}}{\text{Av. no. workers}} + \frac{\text{Discharges}}{\text{Av. no. workers}} + \frac{\text{Laid-off}}{\text{Av. no. workers}}$$

Labour turnover may be analysed in many ways, e.g. between preventable and non-preventable turnover, or between old employees and new employees.

USE OF STATISTICS IN MANAGEMENT

INTERNAL AUDIT

In the same way as outside auditors check the entries in the books of a firm with the corresponding vouchers, so will staff engaged on an internal audit. The great advantage of an internal audit is that it can be continuous and errors can be put right immediately. It is not necessary to check all entries and vouchers, provided the sample which is checked is truly representative of the whole. Unfortunately, the current practice, both in outside audits and internal audits, is to choose haphazard samples selected according to the whim of the individual auditor, thus leading to biased samples. The sample should be chosen according to the rules of "random sampling" as described in Chapter 3 of this book.

BUSINESS REPORTS

These may be:

(a) *On a specific subject,* e.g. the desirability of installing machine accounting, or, perhaps, an inquiry into the number of strikes in the works in the past year. In this case, the report will follow the usual lines—terms of reference, followed by the body of the report, then the conclusions or findings, followed, if required, by recommendations. Statistical matter often forms part of such reports, e.g. tables, charts and graphs.

(b) *Periodical reports on certain aspects of the business,* e.g. sales, production, advertising, etc. A prominent part of such reports will consist of relevant tabulations with appropriate commentary. These reports form the basis for action and policy decisions. Management ratios will often be shown in such reports.

PURCHASING DEPARTMENT

Graphs showing stocks of each commodity where the stock held is large, or graphs showing stocks of groups of commodities where each individual item in the group is stocked only in small quantities, will indicate movements in stocks held. When separate commodities are charted, the graphs will show quantities; when groups of commodities are charted, value will have to be shown. On the graphs, distinctive lines will show "ideal" quantity, maximum quantity and minimum quantity or re-ordering quantity. These graphs are extremely useful to prevent overbuying.

Stock turnover figures will indicate if the "right" goods are being purchased.

Graphs of prices over a period of years may indicate the most favourable time of year to purchase, but this may conflict with other considerations: it costs money to hold stocks.

DEBT-COLLECTION MANAGEMENT

A chart of the debtors' turnover will indicate if debts are taking longer to collect. This may be due to accepting less credit-worthy customers or changing terms of sale or less efficient debt management, or possibly changing general economic conditions.

ADVERTISING DEPARTMENT

Inquiries and orders should be shown on the same semi-logarithmic chart. Differences between rates of change may indicate inefficient (or otherwise) handling of inquiries. Similarly, advertising costs and sales shown on a semi-logarithmic graph will indicate the effectiveness of advertising. In both cases, there will be a time lag to take into consideration.

SALES DEPARTMENT

Tables and/or graphs showing the following information have obvious uses:

(a) comparison of actual and estimated sales;

(b) analysis of sales according to method of sale, e.g. mail order, salesmen, dealers, etc.;

(c) analysis of sales according to area, commodity, credit or cash, etc.;

(d) analysis of sales expenses, advertising, salaries, commission, etc.;

(e) analysis of the performance of salesmen, showing calls made, orders obtained, ratio of orders to calls and average cost per order in respect of each salesman.

PRODUCTION DEPARTMENT

Graphs showing production and sales will indicate whether to increase or decrease production. Graphs showing output and labour costs indicate the efficiency of labour. These graphs should be drawn on semi-logarithmic paper.

Among the many charts which are useful mention must be made of Gantt progress charts (*see* Chapter 9) and quality control charts (*see* Chapter 23).

PERSONNEL DEPARTMENT

Among the matters that may be dealt with by means of tables and charts, the following may be cited:

(a) accidents and injuries, analysed according to seriousness, department and cause;

(b) progress of trainees;

(c) absenteeism analysed according to reason and department;

(d) labour turnover.

QUESTIONS

1. Of what use are management ratios? Illustrate your answer by means of examples.

2. You are required to prepare the sales report for your firm. Mention some of the statistical material which will form part of such a report.

3. Explain how time series analysis can be used to prepare the sales budget in budgetary control. What precautions are necessary in fixing the sales budget?

4. A manager of a retail distribution firm asks you for advice on the most appropriate statistical methods for use in his business. Indicate in a short report what you feel to be the most satisfactory statistical methods for aiding management, the tasks which they perform, and their contribution to better efficiency.

APPENDIX A

Logarithm Tables

	0	1	2	3	4	5	6	7	8	9	1	2	3	4	5	6	7	8	9
10	0000	0043	0086	0128	0170						5	9	13	17	21	26	30	34	38
						0212	0253	0294	0334	0374	4	8	12	16	20	24	28	32	36
11	0414	0453	0492	0531	0569						4	8	12	16	20	23	27	31	35
						0607	0645	0682	0719	0755	4	7	11	15	18	22	26	29	33
12	0792	0828	0864	0899	0934						3	7	11	14	18	21	25	28	32
						0969	1004	1038	1072	1106	3	7	10	14	17	20	24	27	31
13	1139	1173	1206	1239	1271						3	6	10	13	16	19	23	26	29
						1303	1335	1367	1399	1430	3	7	10	13	16	19	22	25	29
14	1461	1492	1523	1553	1584						3	6	9	12	15	19	22	25	28
						1614	1644	1673	1703	1732	3	6	9	12	14	17	20	23	26
15	1761	1790	1818	1847	1875						3	6	9	11	14	17	20	23	26
						1903	1931	1959	1987	2014	3	6	8	11	14	17	19	22	25
16	2041	2068	2095	2122	2148						3	6	8	11	14	16	19	22	24
						2175	2201	2227	2253	2279	3	5	8	10	13	16	18	21	23
17	2304	2330	2355	2380	2405						3	5	8	10	13	15	18	20	23
						2430	2455	2480	2504	2529	3	5	8	10	12	15	17	20	22
18	2553	2577	2601	2625	2648						2	5	7	9	12	14	17	19	21
						2672	2695	2718	2742	2765	2	4	7	9	11	14	16	18	21
19	2788	2810	2833	2856	2878						2	4	7	9	11	13	16	18	20
						2900	2923	2945	2967	2989	2	4	6	8	11	13	15	17	19
20	3010	3032	3054	3075	3096	3118	3139	3160	3181	3201	2	4	6	8	11	13	15	17	19
21	3222	3243	3263	3284	3304	3324	3345	3365	3385	3404	2	4	6	8	10	12	14	16	18
22	3424	3444	3464	3483	3502	3522	3541	3560	3579	3598	2	4	6	8	10	12	14	15	17
23	3617	3636	3655	3674	3692	3711	3729	3747	3766	3784	2	4	6	7	9	11	13	15	17
24	3802	3820	3838	3856	3874	3892	3909	3927	3945	3962	2	4	5	7	9	11	12	14	16
25	3979	3997	4014	4031	4048	4065	4082	4099	4116	4133	2	3	5	7	9	10	12	14	15
26	4150	4166	4183	4200	4216	4232	4249	4265	4281	4298	2	3	5	7	8	10	11	13	15
27	4314	4330	4346	4362	4378	4393	4409	4425	4440	4456	2	3	5	6	8	9	11	13	14
28	4472	4487	4502	4518	4533	4548	4564	4579	4594	4609	2	3	5	6	8	9	11	12	14
29	4624	4639	4654	4669	4683	4698	4713	4728	4742	4757	1	3	4	6	7	9	10	12	13
30	4771	4786	4800	4814	4829	4843	4857	4871	4886	4900	1	3	4	6	7	9	10	11	13
31	4914	4928	4942	4955	4969	4983	4997	5011	5024	5038	1	3	4	6	7	8	10	11	12
32	5051	5065	5079	5092	5105	5119	5132	5145	5159	5172	1	3	4	5	7	8	9	11	12
33	5185	5198	5211	5224	5237	5250	5263	5276	5289	5302	1	3	4	5	6	8	9	10	12
34	5315	5328	5340	5353	5366	5378	5391	5403	5416	5428	1	3	4	5	6	8	9	10	11
35	5441	5453	5465	5478	5490	5502	5514	5527	5539	5551	1	2	4	5	6	7	9	10	11
36	5563	5575	5587	5599	5611	5623	5635	5647	5658	5670	1	2	4	5	6	7	8	10	11
37	5682	5694	5705	5717	5729	5740	5752	5763	5775	5786	1	2	3	5	6	7	8	9	10
38	5798	5809	5821	5832	5843	5855	5866	5877	5888	5899	1	2	3	5	6	7	8	9	10
39	5911	5922	5933	5944	5955	5966	5977	5988	5999	6010	1	2	3	4	5	7	8	9	10
40	6021	6031	6042	6053	6064	6075	6085	6096	6107	6117	1	2	3	4	5	6	8	9	10
41	6128	6138	6149	6160	6170	6180	6191	6201	6212	6222	1	2	3	4	5	6	7	8	9
42	6232	6243	6253	6263	6274	6284	6294	6304	6314	6325	1	2	3	4	5	6	7	8	9
43	6335	6345	6355	6365	6375	6385	6395	6405	6415	6425	1	2	3	4	5	6	7	8	9
44	6435	6444	6454	6464	6474	6484	6493	6503	6513	6522	1	2	3	4	5	6	7	8	9
45	6532	6542	6551	6561	6571	6580	6590	6599	6609	6618	1	2	3	4	5	6	7	8	9
46	6628	6637	6646	6656	6665	6675	6684	6693	6702	6712	1	2	3	4	5	6	7	7	8
47	6721	6730	6739	6749	6758	6767	6776	6785	6794	6803	1	2	3	4	5	5	6	7	8
48	6812	6821	6830	6839	6848	6857	6866	6875	6884	6893	1	2	3	4	4	5	6	7	8
49	6902	6911	6920	6928	6937	6946	6955	6964	6972	6981	1	2	3	4	4	5	6	7	8

	0	1	2	3	4	5	6	7	8	9	1	2	3	4	5	6	7	8	9
50	6990	6998	7007	7016	7024	7033	7042	7050	7059	7067	1	2	3	3	4	5	6	7	8
51	7076	7084	7093	7101	7110	7118	7126	7135	7143	7152	1	2	3	3	4	5	6	7	8
52	7160	7168	7177	7185	7193	7202	7210	7218	7226	7235	1	2	2	3	4	5	6	7	7
53	7243	7251	7259	7267	7275	7284	7292	7300	7308	7316	1	2	2	3	4	5	6	6	7
54	7324	7332	7340	7348	7356	7364	7372	7380	7388	7396	1	2	2	3	4	5	6	6	7
55	7404	7412	7419	7427	7435	7443	7451	7459	7466	7474	1	2	2	3	4	5	5	6	7
56	7482	7490	7497	7505	7513	7520	7528	7536	7543	7551	1	2	2	3	4	5	5	6	7
57	7559	7566	7574	7582	7589	7597	7604	7612	7619	7627	1	2	2	3	4	5	5	6	7
58	7634	7642	7649	7657	7664	7672	7679	7686	7694	7701	1	1	2	3	4	4	5	6	7
59	7709	7716	7723	7731	7738	7745	7752	7760	7767	7774	1	1	2	3	4	4	5	6	7
60	7782	7789	7796	7803	7810	7818	7825	7832	7839	7846	1	1	2	3	4	4	5	6	6
61	7853	7860	7868	7875	7882	7889	7896	7903	7910	7917	1	1	2	3	4	4	5	6	6
62	7924	7931	7938	7945	7952	7959	7966	7973	7980	7987	1	1	2	3	3	4	5	6	6
63	7993	8000	8007	8014	8021	8028	8035	8041	8048	8055	1	1	2	3	3	4	5	5	6
64	8062	8069	8075	8082	8089	8096	8102	8109	8116	8122	1	1	2	3	3	4	5	5	6
65	8129	8136	8142	8149	8156	8162	8169	8176	8182	8189	1	1	2	3	3	4	5	5	6
66	8195	8202	8209	8215	8222	8228	8235	8241	8248	8254	1	1	2	3	3	4	5	5	6
67	8261	8267	8274	8280	8287	8293	8299	8306	8312	8319	1	1	2	3	3	4	5	5	6
68	8325	8331	8338	8344	8351	8357	8363	8370	8376	8382	1	1	2	3	4	4	5	5	6
69	8388	8395	8401	8407	8414	8420	8426	8432	8439	8445	1	1	2	2	3	4	4	5	6
70	8451	8457	8463	8470	8476	8482	8488	8494	8500	8506	1	1	2	2	3	4	4	5	6
71	8513	8519	8525	8531	8537	8543	8549	8555	8561	8567	1	1	2	2	3	4	4	5	5
72	8573	8579	8585	8591	8597	8603	8609	8615	8621	8627	1	1	2	2	3	4	4	5	5
73	8633	8639	8645	8651	8657	8663	8669	8675	8681	8686	1	1	2	2	3	4	4	5	5
74	8692	8698	8704	8710	8716	8722	8727	8733	8739	8745	1	1	2	2	3	4	4	5	5
75	8751	8756	8762	8768	8774	8779	8785	8791	8797	8802	1	1	2	2	3	3	4	5	5
76	8808	8814	8820	8825	8831	8837	8842	8848	8854	8859	1	1	2	2	3	3	4	5	5
77	8865	8871	8876	8882	8887	8893	8899	8904	8910	8915	1	1	2	2	3	3	4	4	5
78	8921	8927	8932	8938	8943	8949	8954	8960	8965	8971	1	1	2	2	3	3	4	4	5
79	8976	8982	8987	8993	8998	9004	9009	9015	9020	9025	1	1	2	2	3	3	4	4	5
80	9031	9036	9042	9047	9053	9058	9063	9069	9074	9079	1	1	2	2	3	3	4	4	5
81	9085	9090	9096	9101	9106	9112	9117	9122	9128	9133	1	1	2	2	3	3	4	4	5
82	9138	9143	9149	9154	9159	9165	9170	9175	9180	9186	1	1	2	2	3	3	4	4	5
83	9191	9196	9201	9206	9212	9217	9222	9227	9232	9238	1	1	2	2	3	3	4	4	5
84	9243	9248	9253	9258	9263	9269	9274	9279	9284	9289	1	1	2	2	3	3	4	4	5
85	9294	9299	9304	9309	9315	9320	9325	9330	9335	9340	1	1	2	2	3	3	4	4	5
86	9345	9350	9355	9360	9365	9370	9375	9380	9385	9390	1	1	2	2	3	3	4	4	5
87	9395	9400	9405	9410	9415	9420	9425	9430	9435	9440	0	1	1	2	2	3	3	4	4
88	9445	9450	9455	9460	9465	9469	9474	9479	9484	9489	0	1	1	2	2	3	3	4	4
89	9494	9499	9504	9509	9513	9518	9523	9528	9533	9538	0	1	1	2	2	3	3	4	4
90	9542	9547	9552	9557	9562	9566	9571	9576	9581	9586	0	1	1	2	2	3	3	4	4
91	9590	9595	9600	9605	9609	9614	9619	9624	9628	9633	0	1	1	2	2	3	3	4	4
92	9638	9643	9647	9652	9657	9661	9666	9671	9675	9680	0	1	1	2	2	3	3	4	4
93	9685	9689	9694	9699	9703	9708	9713	9717	9722	9727	0	1	1	2	2	3	3	4	4
94	9731	9736	9741	9745	9750	9754	9759	9763	9768	9773	0	1	1	2	2	3	3	4	4
95	9777	9782	9786	9791	9795	9800	9805	9809	9814	9818	0	1	1	2	2	3	3	4	4
96	9823	9827	9832	9836	9841	9845	9850	9854	9859	9863	0	1	1	2	2	3	3	4	4
97	9868	9872	9877	9881	9886	9890	9894	9899	9903	9908	0	1	1	2	2	3	3	4	4
98	9912	9917	9921	9926	9930	9934	9939	9943	9948	9952	0	1	1	2	2	3	3	4	4
99	9956	9961	9965	9969	9974	9978	9983	9987	9991	9996	0	1	1	2	2	3	3	3	4

Given the logarithm to find the number: The logarithm must be looked for in the body of the table (the *mantissa* part only).

APPENDIX B

Table of Squares

	0	1	2	3	4	5	6	7	8	9
10	1 00 00	1 02 01	1 04 04	1 06 09	1 08 16	1 10 25	1 12 36	1 14 49	1 16 64	1 18 81
11	1 21 00	1 23 21	1 25 44	1 27 69	1 29 96	1 32 25	1 34 56	1 36 89	1 39 24	1 41 61
12	1 44 00	1 46 41	1 48 84	1 51 29	1 53 76	1 56 25	1 58 76	1 61 29	1 63 84	1 66 41
13	1 69 00	1 71 61	1 74 24	1 76 89	1 79 56	1 82 25	1 84 96	1 87 69	1 90 44	1 93 21
14	1 96 00	1 98 81	2 01 64	2 04 49	2 07 36	2 10 25	2 13 16	2 16 09	2 19 04	2 22 01
15	2 25 00	2 28 01	2 31 04	2 34 09	2 37 16	2 40 25	2 43 36	2 46 49	2 49 64	2 52 81
16	2 56 00	2 59 21	2 62 44	2 65 69	2 68 96	2 72 25	2 75 56	2 78 89	2 82 24	2 85 61
17	2 89 00	2 92 41	2 95 84	2 99 29	3 02 76	3 06 25	3 09 76	3 13 29	3 16 84	3 20 41
18	3 24 00	3 27 61	3 31 24	3 34 89	3 38 56	3 42 25	3 45 96	3 49 69	3 53 44	3 57 21
19	3 61 00	3 64 81	3 68 64	3 72 49	3 76 36	3 80 25	3 84 16	3 88 09	3 92 04	3 96 01
20	4 00 00	4 04 01	4 08 04	4 12 09	4 16 16	4 20 25	4 24 36	4 28 49	4 32 64	4 36 81
21	4 41 00	4 45 21	4 49 44	4 53 69	4 57 96	4 62 25	4 66 56	4 70 89	4 75 24	4 79 61
22	4 84 00	4 88 41	4 92 84	4 97 29	5 01 76	5 06 25	5 10 76	5 15 29	5 19 84	5 24 41
23	5 29 00	5 33 61	5 38 24	5 42 89	5 47 56	5 52 25	5 56 96	5 61 69	5 66 44	5 71 21
24	5 76 00	5 80 81	5 85 64	5 90 49	5 95 36	6 00 25	6 05 16	6 10 09	6 15 04	6 20 01
25	6 25 00	6 30 01	6 35 04	6 40 09	6 45 16	6 50 25	6 55 36	6 60 49	6 65 64	6 70 81
26	6 76 00	6 81 21	6 86 44	6 91 69	6 96 96	7 02 25	7 07 56	7 12 89	7 18 24	7 23 61
27	7 29 00	7 34 41	7 39 84	7 45 29	7 50 76	7 56 25	7 61 76	7 67 29	7 72 84	7 78 41
28	7 84 00	7 89 61	7 95 24	8 00 89	8 06 56	8 12 25	8 17 96	8 23 69	8 29 44	8 35 21
29	8 41 00	8 46 81	8 52 64	8 58 49	8 64 36	8 70 25	8 76 16	8 82 09	8 88 04	8 94 01
30	9 00 00	9 06 01	9 12 04	9 18 09	9 24 16	9 30 25	9 36 36	9 42 49	9 48 64	9 54 81
31	9 61 00	9 67 21	9 73 44	9 79 69	9 85 96	9 92 25	9 98 56	10 04 89	10 11 24	10 17 61
32	10 24 00	10 30 41	10 36 84	10 43 29	10 49 76	10 56 25	10 62 76	10 69 29	10 75 84	10 82 41
33	10 89 00	10 95 61	11 02 24	11 08 89	11 15 56	11 22 25	11 28 96	11 35 69	11 42 44	11 49 21
34	11 56 00	11 62 81	11 69 64	11 76 49	11 83 36	11 90 25	11 97 16	12 04 09	12 11 04	12 18 01
35	12 25 00	12 32 01	12 39 04	12 46 09	12 53 16	12 60 25	12 67 36	12 74 49	12 81 64	12 88 81
36	12 96 00	13 03 21	13 10 44	13 17 69	13 24 96	13 32 25	13 39 56	13 46 89	13 54 24	13 61 61
37	13 69 00	13 76 41	13 83 84	13 91 29	13 98 76	14 06 25	14 13 76	14 21 29	14 28 84	14 36 41
38	14 44 00	14 51 61	14 59 24	14 66 89	14 74 56	14 82 25	14 89 96	14 97 69	15 05 44	15 13 21
39	15 21 00	15 28 81	15 36 64	15 44 49	15 52 36	15 60 25	15 68 16	15 76 09	15 84 04	15 92 01
40	16 00 00	16 08 01	16 16 04	16 24 09	16 32 16	16 40 25	16 48 36	16 56 49	16 64 64	16 72 81
41	16 81 00	16 89 21	16 97 44	17 05 69	17 13 96	17 22 25	17 30 56	17 38 89	17 47 24	17 55 61
42	17 64 00	17 72 41	17 80 84	17 89 29	17 97 76	18 06 25	18 14 76	18 23 29	I8 31 84	18 40 41
43	18 49 00	18 57 61	18 66 24	18 74 89	18 83 56	18 92 25	19 00 96	19 09 69	19 18 44	19 27 21
44	19 36 00	19 44 81	19 53 64	19 62 49	19 71 36	19 80 25	19 89 16	19 98 09	20 07 04	20 16 01
45	20 25 00	20 34 01	20 43 04	20 52 09	20 61 16	20 70 25	20 79 36	20 88 49	20 97 64	21 06 81
46	21 16 00	21 25 21	21 34 44	21 43 69	21 52 96	21 62 25	21 71 56	21 80 89	21 90 24	21 99 61
47	22 09 00	22 18 41	22 27 84	22 37 29	22 46 76	22 56 25	22 65 76	22 75 29	22 84 84	22 94 41
48	23 04 00	23 13 61	23 23 24	23 32 89	23 42 56	23 52 25	23 61 96	23 71 69	23 81 44	23 91 21
49	24 01 00	24 10 81	24 20 64	24 30 49	24 40 36	24 50 25	24 60 16	24 70 09	24 80 04	24 90 01
50	25 00 00	25 10 01	25 20 04	25 30 09	25 40 16	25 50 25	25 60 36	25 70 49	25 80 64	25 90 81
51	26 01 00	26 11 21	26 21 44	26 31 69	26 41 96	26 52 25	26 62 56	26 72 89	26 83 24	26 93 61
52	27 04 00	27 14 41	27 24 84	27 35 29	27 45 76	27 56 25	27 66 76	27 77 29	27 87 84	27 98 41
53	28 09 00	28 19 61	28 30 24	28 40 89	28 51 56	28 62 25	28 72 96	28 83 69	28 94 44	29 05 21
54	29 16 00	29 26 81	29 37 64	29 48 49	29 59 36	29 70 25	29 81 16	29 92 09	30 03 04	30 14 01

	0	1	2	3	4	5	6	7	8	9
55	30 25 00	30 36 01	30 47 04	30 58 09	30 69 16	30 80 25	30 91 36	31 02 49	31 13 64	31 24 81
56	31 36 00	31 47 21	31 58 44	31 69 69	31 80 96	31 92 25	32 03 56	32 14 89	32 26 24	32 37 61
57	32 49 00	32 60 41	32 71 84	32 83 29	32 94 76	33 06 25	33 17 76	33 29 29	33 40 84	33 52 41
58	33 64 00	33 75 61	33 87 24	33 98 89	34 10 56	34 22 25	34 33 96	34 45 69	34 57 44	34 69 21
59	34 81 00	34 92 81	35 04 64	35 16 49	35 28 36	35 40 25	35 52 16	35 64 09	35 76 04	35 88 01
60	36 00 00	36 12 01	36 24 04	36 36 09	36 48 16	36 60 25	36 72 36	36 84 49	36 96 64	37 08 81
61	37 21 00	37 33 21	37 45 44	37 57 69	37 69 96	37 82 25	37 94 56	38 06 89	38 19 24	38 31 61
62	38 44 00	38 56 41	38 68 84	38 81 29	38 93 76	39 06 25	39 18 76	39 31 29	39 43 84	39 56 41
63	39 69 00	39 81 61	39 94 24	40 06 89	40 19 56	40 32 25	40 44 96	40 57 69	40 70 44	40 83 21
64	40 96 00	41 08 81	41 21 64	41 34 49	41 47 36	41 60 25	41 73 16	41 86 09	41 99 04	42 12 01
65	42 25 00	42 38 01	42 51 04	42 64 09	42 77 16	42 90 25	43 03 36	43 16 49	43 29 64	43 42 81
66	43 56 00	43 69 21	43 82 44	43 95 69	44 08 96	44 22 25	44 35 56	44 48 89	44 62 24	44 75 61
67	44 89 00	45 02 41	45 15 84	45 29 29	45 42 76	45 56 25	45 69 76	45 83 29	45 96 84	46 10 41
68	46 24 00	46 37 61	46 51 24	46 64 89	46 78 56	46 92 25	47 05 96	47 19 69	47 33 44	47 47 21
69	47 61 00	47 74 81	47 88 64	48 02 49	48 16 36	48 30 25	48 44 16	48 58 09	48 72 04	48 86 01
70	49 00 00	49 14 01	49 28 04	49 42 09	49 56 16	49 70 25	49 84 36	49 98 49	50 12 64	50 26 81
71	50 41 00	50 55 21	50 69 44	50 83 69	50 97 96	51 12 25	51 26 56	51 40 89	51 55 24	51 69 61
72	51 84 00	51 98 41	52 12 84	52 27 29	52 41 76	52 56 25	52 70 76	52 85 29	52 99 84	53 14 41
73	53 29 00	53 43 61	53 58 24	53 72 89	53 87 56	54 02 25	54 16 96	54 31 69	54 46 44	54 61 21
74	54 76 00	54 90 81	55 05 64	55 20 49	55 35 36	55 50 25	55 65 16	55 80 09	55 95 04	56 10 01
75	56 25 00	56 40 01	56 55 04	56 70 09	56 85 16	57 00 25	57 15 36	57 30 49	57 45 64	57 60 81
76	57 76 00	57 91 21	58 06 44	58 21 69	58 36 96	58 52 25	58 67 56	58 82 89	58 98 24	59 13 61
77	59 29 00	59 44 41	59 59 84	59 75 29	59 90 76	60 06 25	60 21 76	60 37 29	60 52 84	60 68 41
78	60 84 00	60 99 61	61 15 24	61 30 89	61 46 56	61 62 25	61 77 96	61 93 69	62 09 44	62 25 21
79	62 41 00	62 56 81	62 72 64	62 88 49	63 04 36	63 20 25	63 36 16	63 52 09	63 68 04	63 84 01
80	64 00 00	64 16 01	64 32 04	64 48 09	64 64 16	64 80 25	64 96 36	65 12 49	65 28 64	65 44 81
81	65 61 00	65 77 21	65 93 44	66 09 69	66 25 96	66 42 25	66 58 56	66 74 89	66 91 24	67 07 61
82	67 24 00	67 40 41	67 56 84	67 73 29	67 89 76	68 06 25	68 22 76	68 39 29	68 55 84	68 72 41
83	68 89 00	69 05 61	69 22 24	69 38 89	69 55 56	69 72 25	69 88 96	70 05 69	70 22 44	70 39 21
84	70 56 00	70 72 81	70 89 64	71 06 49	71 23 36	71 40 25	71 57 16	71 74 09	71 91 04	72 08 01
85	72 25 00	72 42 01	72 59 04	72 76 09	72 93 16	73 10 25	73 27 36	73 44 49	73 61 64	73 78 81
86	73 96 00	74 13 21	74 30 44	74 47 69	74 64 96	74 82 25	74 99 56	75 16 89	75 34 24	75 51 61
87	75 69 00	75 86 41	76 03 84	76 21 29	76 38 76	76 56 25	76 73 76	76 91 29	77 08 84	77 26 41
88	77 44 00	77 61 61	77 79 24	77 96 89	78 14 56	78 32 25	78 49 96	78 67 69	78 85 44	79 03 21
89	79 21 00	79 38 81	79 56 64	79 74 49	79 92 36	80 10 25	80 28 16	80 46 09	80 64 04	80 82 01
90	81 00 00	81 18 01	81 36 04	81 54 09	81 72 16	81 90 25	82 08 36	82 26 49	82 44 64	82 62 81
91	82 81 00	82 99 21	83 17 44	83 35 69	83 53 96	83 72 25	83 90 56	84 08 89	84 27 24	84 45 61
92	84 64 00	84 82 41	85 00 84	85 19 29	85 37 76	85 56 25	85 74 76	85 93 29	86 11 84	86 30 41
93	86 49 00	86 67 61	86 86 24	87 04 89	87 23 56	87 42 25	87 60 96	87 79 69	87 98 44	88 17 21
94	88 36 00	88 54 81	88 73 64	88 92 49	89 11 36	89 30 25	89 49 16	89 68 09	89 87 04	90 06 01
95	90 25 00	90 44 01	90 63 04	90 82 09	91 01 16	91 20 25	91 39 36	91 58 49	91 77 64	91 96 81
96	92 16 00	92 35 21	92 54 44	92 73 69	92 92 96	93 12 25	93 31 56	93 50 89	93 70 24	93 89 61
97	94 09 00	94 28 41	94 47 84	94 67 29	94 86 76	95 06 25	95 25 76	95 45 29	95 64 84	95 84 41
98	96 04 00	96 23 61	96 43 24	96 62 89	96 82 56	97 02 25	97 21 96	97 41 69	97 61 44	97 81 21
99	98 01 00	98 20 81	98 40 64	98 60 49	98 80 36	99 00 25	99 20 16	99 40 09	99 60 04	99 80 01

Method of using table. The first two figures of the number to be squared are found in the first column. The third figure of the number to be squared is found along the top row numbered 0 to 9.

Example. Find the square of 30.7.

The figures 30 will be found in the first column. This will give the row. The figure found in *this* row under column 7 is 94249. This is the square of 307; therefore the square of 30.7 is 942.49.

If the square of a two-figure number *only* is required, the result is given in the column headed "0" *but* the last 2 noughts must be ignored.

APPENDIX C

Area Under Normal Curve

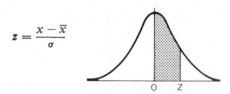

$$z = \frac{x - \overline{x}}{\sigma}$$

z	0	1	2	3	4	5	6	7	8	9
0·0	0·0000	0·0040	0·0080	0·0120	0·0160	0·0199	0·0239	0·0279	0·0319	0·0359
0·1	0·0398	0·0438	0·0478	0·0517	0·0557	0·0596	0·0636	0·0675	0·0714	0·0754
0·2	0·0793	0·0832	0·0871	0·0910	0·0948	0·0987	0·1026	0·1064	0·1103	0·1141
0·3	0·1179	0·1217	0·1255	0·1293	0·1331	0·1368	0·1406	0·1443	0·1480	0·1517
0·4	0·1554	0·1591	0·1628	0·1664	0·1700	0·1736	0·1772	0·1808	0·1844	0·1879
0·5	0·1915	0·1950	0·1985	0·2019	0·2054	0·2088	0·2123	0·2157	0·2190	0·2224
0·6	0·2258	0·2291	0·2324	0·2357	0·2389	0·2422	0·2454	0·2486	0·2518	0·2549
0·7	0·2580	0·2612	0·2642	0·2673	0·2704	0·2734	0·2764	0·2794	0·2823	0·2852
0·8	0·2881	0·2910	0·2939	0·2967	0·2996	0·3023	0·3051	0·3078	0·3106	0·3133
0·9	0·3159	0·3186	0·3212	0·3238	0·3264	0·3289	0·3315	0·3340	0·3365	0·3389
1·0	0·3413	0·3438	0·3461	0·3485	0·3508	0·3531	0·3554	0·3577	0·3599	0·3621
1·1	0·3643	0·3665	0·3686	0·3708	0·3729	0·3749	0·3770	0·3790	0·3810	0·3830
1·2	0·3849	0·3869	0·3888	0·3907	0·3925	0·3944	0·3962	0·3980	0·3997	0·4015
1·3	0·4032	0·4049	0·4066	0·4082	0·4099	0·4115	0·4131	0·4147	0·4162	0·4177
1·4	0·4192	0·4207	0·4222	0·4236	0·4251	0·4265	0·4279	0·4292	0·4306	0·4319
1·5	0·4332	0·4345	0·4357	0·4370	0·4382	0·4394	0·4406	0·4418	0·4429	0·4441
1·6	0·4452	0·4463	0·4474	0·4484	0·4495	0·4505	0·4515	0·4525	0·4535	0·4545
1·7	0·4554	0·4564	0·4573	0·4582	0·4591	0·4599	0·4608	0·4616	0·4625	0·4633
1·8	0·4641	0·4649	0·4656	0·4664	0·4671	0·4678	0·4686	0·4693	0·4699	0·4706
1·9	0·4713	0·4719	0·4726	0·4732	0·4738	0·4744	0·4750	0·4756	0·4761	0·4767
2·0	0·4772	0·4778	0·4783	0·4788	0·4793	0·4798	0·4803	0·4808	0·4812	0·4817
2·1	0·4821	0·4826	0·4830	0·4834	0·4838	0·4842	0·4846	0·4850	0·4854	0·4857
2·2	0·4861	0·4864	0·4868	0·4871	0·4875	0·4878	0·4881	0·4884	0·4887	0·4890
2·3	0·4893	0·4896	0·4898	0·4901	0·4904	0·4906	0·4909	0·4911	0·4913	0·4916
2·4	0·4918	0·4920	0·4922	0·4925	0·4927	0·4929	0·4931	0·4932	0·4934	0·4936
2·5	0·4938	0·4940	0·4941	0·4943	0·4945	0·4946	0·4948	0·4949	0·4951	0·4952
2·6	0·4953	0·4955	0·4956	0·4957	0·4959	0·4960	0·4961	0·4962	0·4963	0·4964
2·7	0·4965	0·4966	0·4967	0·4968	0·4969	0·4970	0·4971	0·4972	0·4973	0·4974
2·8	0·4974	0·4975	0·4976	0·4977	0·4977	0·4978	0·4979	0·4979	0·4980	0·4981
2·9	0·4981	0·4982	0·4982	0·4983	0·4984	0·4984	0·4985	0·4985	0·4986	0·4986
3·0	0·4987	0·4987	0·4987	0·4988	0·4988	0·4989	0·4989	0·4989	0·4990	0·4990

The t Distribution

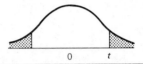

Number of degrees of freedom	Probability			
	10%	5%	2%	1%
1	6.314	12.706	31.821	63.657
2	2.910	4.303	6.965	9.925
3	2.353	3.182	4.541	5.841
4	2.132	2.776	3.747	4.604
5	2.015	2.571	3.365	4.032
6	1.943	2.447	3.143	3.707
7	1.895	2.365	2.998	3.499
8	1.860	2.306	2.896	3.355
9	1.833	2.262	2.821	3.250
10	1.812	2.228	2.764	3.169
11	1.796	2.201	2.718	3.106
12	1.782	2.179	2.681	3.055
13	1.771	2.160	2.650	3.012
14	1.761	2.145	2.624	2.977
15	1.753	2.131	2.602	2.947
16	1.746	2.120	2.583	2.921
17	1.740	2.110	2.567	2.898
18	1.734	2.101	2.552	2.878
19	1.729	2.093	2.539	2.861
20	1.725	2.086	2.528	2.845

Note. The t table above gives the probabilities (significance levels) for two tails. The table, however, can be used for one tail tests. The same value of the t statistic gives a 1 per cent level of significance for a one tail test and a 2 per cent for a two tail test. The same value of the t statistic gives a 5 per cent level for a one tail test and a 10 per cent level for a two tail test.

Whereas the areas under the "normal" curve (Appendix C) are in the body of the table, it is the values of the t statistic which are in the body of the t table.

Percentage Points of the χ^2 Distribution

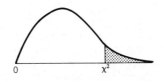

d.f.	30	10	5	2.5	1	0.5	0.1
1	1.07	2.71	3.84	5.02	6.63	7.88	10.83
2	2.41	4.61	5.99	7.38	9.21	10.60	13.81
3	3.67	6.25	7.81	9.35	11.34	12.84	16.27
4	4.88	7.78	9.49	11.14	13.28	14.86	18.47
5	6.06	9.24	11.07	12.83	15.09	16.75	20.52
6	7.23	10.64	12.59	14.45	16.81	18.55	22.46
7	8.38	12.02	14.07	16.01	18.48	20.28	24.32
8	9.52	13.36	15.51	17.53	20.09	21.95	26.12
9	10.66	14.68	16.92	19.02	21.67	23.59	27.88
10	11.78	15.99	18.31	20.48	23.21	25.19	29.59

This is, of course, only part of a table. A more complete table would give other probabilities and for other numbers of degrees of freedom.

From the table it can be seen that with one degree of freedom the probability of getting χ^2 as great as, or greater than 3.84 is 5 per cent, a 1 in 20 chance, and this would probably be considered significant.

Note. The smaller the value of χ^2 (the figures in the body of the table), the greater the probability of obtaining it and the better the goodness of fit; the greater the value of χ^2, the smaller the probability of obtaining it, and the greater the probability that given attributes are related.

Quality Control Factors

Sample size	A Warning	Action	L	M	D
2	1.23	1.94	0.18	0.80	4.12
3	0.67	1.05	0.27	0.77	2.98
4	0.48	0.75	0.33	0.75	2.57
5	0.38	0.59	0.37	0.73	2.34
6	0.32	0.50	0.41	0.71	2.21

APPENDIX G

Value of e^{-m}

m	1	2	3	4	4.8	5	6
e^{-m}	0.36788	0.13534	0.04979	0.01832	0.00823	0.00674	0.00248

This is only an extract from a much larger table. Values of e^{-m} can easily be calculated by means of logarithms.

Example. Evaluate $e^{-5.5}$.

$$e^{-5.5} = \frac{1}{e^{5.5}} = \frac{1}{2.71828^{5.5}} \qquad \text{(e is a constant approximately equal to 2.71828)}$$

No.	Log	Log.Log
2.71828	0.4343	$\bar{1}.6378$
5.5		0.7404
2.71828$^{5.5}$	2.389	0.3782
1	0	
0.00409	$\bar{3}.611$	

$$e^{-5.5} = 0.00409.$$

APPENDIX H

Factorials

x	$x!$	x	$x!$	x	$x!$
1	1	5	120	9	362880
2	2	6	720	10	3628800
3	6	7	5040	11	39916800
4	24	8	40320	12	479001600

$x! = x(x - 1)(x - 2)(x - 3)(x - 4) \ldots (3)(2)(1).$
$4! = 4 \times 3 \times 2 \times 1 = 24$

Revision Questions

The following questions have been taken from or based on recent examination papers of the following bodies:

The Institute of Chartered Secretaries and Administrators,
The Association of Certified Accountants,
The Institute of Cost and Management Accountants,
The London Chamber of Commerce and Industry,
The Royal Society of Arts,
The Institute of Statisticians,
The Institute of Personnel Management,
The Chartered Institute of Transport,
The Chartered Institute of Public Finance and Accountancy,
The University of London.

1. Use the method of least squares to forecast the sales of zinc for 1980 from the following time series:

Sales of zinc by value (£m)

Year	Sales of zinc by value
1975	6.4
1976	6.9
1977	7.3
1978	7.7
1979	8.5

2. (a) What is meant by simple random sampling?

(b) A random sample of 1,000 manufactured items is inspected and 250 are found to contain defects. What is the likely range of the proportion defective in the population of items (use 95 per cent confidence limits)?

3. The following table shows the quarterly sales figures in £000s.

Quarterly sales

Years	I	II	III	IV
1976	81	46	42	76
1977	79	40	31	70
1978	64	31	34	66
1979	71	30	32	—

You are required to:

(a) use moving averages to calculate the trend;

(b) calculate the average seasonal variation;

(c) explain how this information can be used in forecasting.

4. The following table shows the trend of cinema admissions and the growth of television licences in the Sutton Coldfield area during 1950–52:

Year	Quarter	Trend of cinema admissions (thousands)	Television licences per 1,000 population
1950	Third	10,025	24
	Fourth	9,924	37
1951	First	9,814	52
	Second	9,726	64
	Third	9,632	69
	Fourth	9,505	81
1952	First	9,405	98
	Second	9,271	101

Calculate the coefficient of correlation and comment on the result.

5. The following grouped frequency table shows the amounts of mortgages made in a period of six months:

Value of mortgage arranged—6 months ending 31st Dec. 1973

Value of mortgage (in £000s)	Number of mortgages arranged
Under 5	36
5 and up to but not including 10	57
10 ” ” 15	44
15 ” ” 20	21
20 ” ” 25	6

Calculate the mean value of the mortgage and the standard deviation (answers should be given correct to the nearest £1). Use all suitable checks in the working.

6. In a survey of customers' accounts, the following grouped frequency table was obtained dealing with credit made available to clients (in weeks):

Credit available to customers

Credit period provided (weeks)	Number of customers
3	46
4	73
5	117
6	35
7	12
8	7

What is the mean credit period provided? In a similar firm it is found that the mean credit period was 5½ weeks (the standard deviation was 1 week). Is there any significant difference between the mean credit periods provided by the two firms?

7. Define the term "error" in statistics.

It is known that the number 45,000 lies between 44,000 and 46,000 and that the number 72,000 lies between 71,500 and 72,500. Find the absolute error when 45,000 is subtracted from 72,000. Calculate this error as a percentage of the difference.

8.

*Production of paper—weekly averages
(thousand tonnes) newsprint*

Quarters	1	2	3	4
1977	10.1	9.7	10.0	10.8
1978	11.3	11.2	8.2	10.8
1979	11.9	12.1	10.4	12.1

Calculate the trend of newsprint production, together with the seasonal variations.

9. The following data concerning industrial accidents and the absences which result from them involve classification by type of employee:

	Men	Type of employee Women	Juveniles
Absence following accidents:			
Up to one month	26	16	8
One month or longer	14	9	7

Is there any evidence to suggest that the severity of accident is associated with type of employee?

Table of chi-squared

No. of degrees of freedom	5% significance point	1% significance point
2	5.991	9.210
3	7.815	11.345
4	9.488	13.277
5	11.070	15.086
6	12.592	16.812

10. What is the importance of the "normal" distribution for the application of quality control methods?

A chocolate manufacturer makes bars of chocolate which have a mean weight of 115 grams and a standard deviation of 12 grams. The manufacturer advertises that the minimum weight is 105 grams. What proportion of bars is likely to be less than this advertised minimum weight?

The manufacturer subsequently decided to advertise that the minimum weight of a bar will be raised to 110 grams. If the standard deviation remains at 12 grams, what mean weight must be obtained to make sure that only 1 per cent of the bars are less than the advertised minimum weight of 110 grams? Assume that the weights of all bars are distributed "normally".

11. Two commodities are of equal importance. Their prices in 1977 and 1978 are given.

	1977	1978
Commodity A (£ per tonne)	12	15
Commodity B (pence per metre)	5	7

Work out an unweighted price index for 1978 taking 1977 as base. Then work out an unweighted index for 1977 taking 1978 as base. Find the product of the *two* index numbers and criticise your result.

12. Explain what is meant by seasonal variation in a time series. Calculate this measure for the production of cigarette tissue paper from the data given in the table below.

Production of cigarette tissue paper
weekly averages (tonnes)

Quarters	1	2	3	4
1977	144	142	129	142
1978	138	101	86	102
1979	127	118	108	129

13. "Perfect accuracy is very seldom obtained in statistics." Explain the meaning of this statement. The productivity of 2,675,000 hectares is between 38 and 39 tonnes of wheat per hectare. Find the yield of wheat in tonnes.

14. What is meant by the term "the standard error of a proportion observed from a random sample"? In a recent survey, a random sample of adults, including 10,000 persons, provided 62 per cent who thought that cigarette smoking affected health. Calculate the standard error of this proportion, and thus indicate the percentage of adults in the population who took this view, showing a range with 95 per cent confidence limits. What sample size would have been necessary to halve these limits?

15. In order to test the effectiveness of a drying agent in paint, the following experiment was carried out. Each of six samples of material was cut into two halves. One half of each was covered with paint containing the agent and the other half with paint without the agent. Then all twelve halves were left to dry. The time taken to dry was as follows:

Drying time (hours)

	sample number					
	1	2	3	4	5	6
Paint with the agent	3.4	3.8	4.2	4.1	3.5	4.7
Paint without the agent	3.6	3.8	4.3	4.3	3.6	4.7

Required:
Carry out a t-test to determine whether the drying agent is effective, giving your reasons for choosing a one tailed or two tailed test. Carefully explain your conclusions.

Selected values of $t_{0.05}$

			Degrees of freedom						
4	*5*	*6*	*7*	*8*	*9*	*10*	*11*	*12*	*13*
One tailed test 2.13	2.02	1.94	1.90	1.86	1.83	1.81	1.80	1.78	1.77
Two tailed test 2.78	2.57	2.45	2.36	2.31	2.26	2.23	2.20	2.18	2.16

16. The breaking strength of 80 test pieces of a certain alloy is given in the following table, the unit being given to the nearest thousand kilograms per square centimetre:

Breaking strength	44–46	46–48	48–50	50–52	52–54	Total
No. of pieces	3	24	27	21	5	80

Calculate the average breaking strength of the alloy and the standard deviation. The numbers are distributed uniformly over the interval in which they lie. Using this fact, calculate the percentage of observations lying within the limits of mean $\pm\sigma$, mean $\pm2\sigma$, mean $\pm3\sigma$, where σ stands for the standard deviation.

17. The following data show the ranking of 10 bloxite producing factories in descending order of safety consciousness, together with the number of accidents per thousand employees over the last five years.

Factory	Rank of safety consciousness	Number of accidents per 1,000 employees
A	1	1.2
B	2	5.3
C	3	8.3
D	4	13.1
E	5	18.2
F	6	15.9
G	7	10.4
H	8	3.3
I	9	23.2
J	10	19.7

(a) By graphing the data, comment on whether accident incidence is associated with safety consciousness.

(b) Calculate Spearman's rank correlation coefficient, R, where:

$$R = 1 - \frac{6\Sigma d^2}{n(n^2 - 1)} \text{, and interpret the result.}$$

(c) Now suppose that it is known that factory H is not typical, being much more modern than the others. Without carrying out the calculations, explain what modifications to your approach in (b) above would be required, and in what way the result is likely to be affected.

18. (a) A manufacturer estimates that the proportion defective in one of his standard production lines is 5 per cent. He takes a random sample of 100 items and finds that 8 are defective—what probability can be ascribed to the sample? Can we infer that the proportion defective of 5 per cent is justified on the basis of this sample?

(b) In City A, a random sample of males reveals that 75 per cent are non-smokers; 1,000 men were questioned to make up the sample. In City B, a similar random sample of 1,000 men gave a proportion of non-smokers as 79 per cent. Can we conclude that there is a significant difference between the two cities in the proportion of non-smoking men?

19. The table below relates to the weekly pay (before tax and other deductions) of the manual wage-earners on a company's pay-roll:

| | April 1970 | | April 1971 | |
	Numbers	Total pay (£)	Numbers	Total pay (£)
Men, aged 21 and over	350	2,500	300	4,200
Women, aged 18 and over	400	1,600	1,200	8,000
Youths and boys	150	450	100	560
Girls	100	250	400	1,540
	1,000	4,800	2,000	14,300

You are required to construct an index of weekly earnings based on 1970 showing the rise of earnings for all employees as one figure.

20. (a) Define the standard error of a mean of samples of n observations and use the concept to explain briefly why one has more confidence in the mean of a large sample than in the mean of a small one.

(b) Show why it is unlikely that a sample of 64 observations with a mean of 5.2 was drawn from a population with a mean of 5.5 and a standard deviation of 0.8.

(c) An expensive piece of equipment is known to be utilised for about 20 per cent of the time.

In order to obtain a more accurate assessment of its utilisation the management decides to carry out some activity sampling, by observing the equipment at random instants of time and noting the percentage of occasions on which it is working.

Determine how many observations will be required to be able to state the utilisation of the equipment within ±5 per cent with 95 per cent certainty.

21. The following data are available concerning the products of a firm taken at random from four factories, and classified according to level of quality:

Output from four factories

Levels of Quality	Factory A	Factory B	Factory C	Factory D
		(Number of items)		
Low	9	12	6	3
Average	22	52	30	16
Good	69	136	64	81

Is there any evidence to suggest that the level of quality is significantly associated with particular factories?

Table of chi-squared

Number of degrees of freedom	5% significance point	1% significance point
3	7.815	11.345
4	9.488	13.277
5	11.070	15.086
6	12.592	16.812
7	14.067	18.475
8	15.507	20.090

22. The following figures, relating to a homogeneous product, were compiled from a company's records:

Number of days taken to complete order	Number of orders
0—4	10
5—9	48
10—14	61
15—19	73
20—24	65
25—29	49
30—34	37
35—39	21
40—44	12
45—49	8
50—54	7
55—59	6
60—64	5
65 and over	10

Calculate: *(a)* the median; *(b)* the two quartiles.

How will the results of your calculations help in determining the delivery dates to be quoted on future orders?

23. Briefly explain how the χ^2 (chi-squared) test may be used to decide whether the two characteristics of classification in a contingency table are associated.

Do the following data give any indication that attending a co-educational school is related to degree examination results?

| Type of school | Degree results | | |
	Successful	Failed	Total
Single sex	863	55	918
Co-educational	360	25	385
Total	1,223	80	1,303

Source: Barnett and Lewis, *Journal of the Royal Statistical Society*, 1963.

24. State briefly what you understand by "skewness". For the distribution below calculate the mean, the standard deviation and the median, hence estimate the degree of skewness.

Steel rings sold

Diameter (cm)	Number (gross)
14.00–14.49	90
14.50–14.99	180
15.00–15.49	200
15.50–15.99	250
16.00–16.49	120
16.50–16.99	80
17.00–17.49	60
17.50–17.99	20

25. In each of the following the degree of approximation in the data is indicated. Obtain in each case the answer required giving the limits of error in the result.

(a) Output: 257 million tonnes; number employed (000) 1,032. Find the output per person.

(b) Hectares (000) 1,766; yield per hectare 19 tonnes. Calculate total output.

26. The average number of employees in a certain firm (excluding staff) was 6,347 in 1953, of whom 43 per cent were absent from their employment for some period during the year. If absences for other than medical reasons are excluded the percentage would be 27. Absence for medical reasons included 187 persons who were absent for periods exceeding two weeks, of whom 15 per cent were absent for more than one month. The number of administrative staff was 628. The exact figure of absentees is not available but is believed to be about 240. There were 176 absentees for duly certified medical reasons, of whom 11 per cent were absent for more than two weeks but less than one month, and 5 persons who were away for periods exceeding one month.

Tabulate these data in a form suitable for presentation in a report to the Board of Directors.

27. The total amount paid in wages by a certain company during 1955 was £W. Let p_i be the per cent of W received by n_i employees who earned £x_i each during the year, so that $p_i = 100 n_i x_i / W$. With this notation the distribution of earnings was as follows:

x(£)	250–	350–	450–	550–	650–	750–	850–1,000
p(%)	14	22	29	16	9	6	4

Find the arithmetic and geometric mean earnings of all employees, and estimate what proportion of employees earned more than £10 a week.

28. The table below shows the number of crimes committed in a sample of local authority areas during a certain period.

Population of local authority (10,000s)	Number of crimes committed
5	12
7	9
8	14
11	20
13	13
15	14
16	22
19	21
22	29
23	28

For the above data, calculate: (a) the product moment correlation coefficient, and (b) a rank correlation coefficient. Indicate how you would interpret these coefficients and which you would regard as the more reliable statistic for purposes of interpreting these data.

29. The following data refers to the value of a broker's deals in securities over 200 days:

Values of deals per day (£s)	Number of days
5,000 and up to, but less than 10,000	47
10,000 ” ” 15,000	63
15,000 ” ” 20,000	71
20,000 ” ” 25,000	12
25,000 ” ” 30,000	7
Total number of days:	200

(a) Calculate the median value of daily sales, and the two quartile values.

(b) Draw a cumulative frequency curve from the data given, and check the values obtained in (a).

30. A commercial company trains its clerical recruits in the completion of forms for entry into the computer system used by the organisation. New employees are allowed to begin work in this area only after they have shown a degree of competence in form

completion, and thus different employees are subject to different training periods. After the completion of training, a check is maintained on errors, and as a result of this the following table is available.

Employee	Number of hours training	Number of errors per 1,000 documents following training
A	12	15
B	17	16
C	28	21
D	20	17
E	18	16

Calculate the product-moment correlation coefficient to determine whether there is significant linkage between training time and error in actual form completion.

Index